Using Meteorology Probability Forecasts in Operational Hydrology

THOMAS E. CROLEY II

Great Lakes Environmental Research Laboratory
National Oceanic and Atmospheric Administration
United States Department of Commerce

American Society of Civil Engineers
1801 Alexander Bell Drive
Reston, Virginia 20191–4400

Abstract: This book presents theory, procedures, and examples for using short-term, seasonal, and interannual forecasts of meteorology probabilities, available every day from the National Oceanic and Atmospheric Administration, Environment Canada, other agencies, and the user. The heuristic approach simultaneously uses forecasts over different time scales, time periods, spatial domains, probability statements, and meteorology variables. The book describes the generation of consequent hydrology (or other) probability forecasts, via operational hydrology methods, for assessing decision risks associated with uncertain meteorology. Any hydrology (or other) model may be used, as illustrated herein. Freely available graphical user interface software is documented to use intuitively and easily the multitude of probabilistic meteorology outlooks available.

Library of Congress Cataloging-in-Publication Data

Croley, Thomas E.
    Using meteorology probability forecasts in operational hydrology / Thomas E. Croley II.
       p.   cm.
    Includes bibliographical references (p.).
    ISBN 0-7844-0459-3
    1. Hydrological forecasting. 2. Probability forecasts (Meteorology). I. Title.

GB845 .C76 2000
551.48'01'12—dc21

                                         99-058022

TABLE OF CONTENTS

ACKNOWLEDGMENTS

The writer appreciates the reviews of this work and the helpful suggestions for its improvement from Timothy Hunter, Brent Lofgren, Peter Landrum, Richard Mickey, Cathy Darnell, and Frank Quinn, among others. Thanks also go to Lisa Ehmer, Joy Chau, and Charlotte McNaughton of ASCE, for their encouragement and guidance. Finally, the writer gratefully thanks his wife Carolyn and his son Ricky for their support. This is Great Lakes Environmental Research Laboratory Contribution No. 1141.

PREFACE

This book describes how to use the numerous forecasts of meteorology probabilities that are now available to the geophysical scientist or engineer. It defines an approach to generating consequent hydrology (or other) probabilities for a designer or manager in assessing decision risks related to *meteorology uncertainty*. This is a heuristic approach that is suitable for simultaneously using a wide variety of disparate probabilistic meteorology forecasts over a variety of time scales, time periods, spatial domains, probability statements, and meteorology variables. The book includes application examples largely from the US Laurentian Great Lakes, but the methodology is not particular to the Great Lakes; readers may use any hydrology (or other *derivative*) model of their own, as the book illustrates. (In fact, readers *must* supply their own models for their applications.) The examples not only include different (hydrology) models, but also illustrate the use of six currently available agency meteorology forecasts in the United States and Canada as well as user-defined meteorology probabilities. The latter include El Niño and La Niña conditional probabilities as well as examples of their derivation and sufficient information for the reader's own applications.

Forecasts of meteorology probabilities are generally available as spatial maps over the United States, Canada, and other countries. This book presents methods to use these forecasts at any specific site on such a map, with the reader's relevant models for that site, to make forecasts of *derivative* probabilities at that site. It is not a treatise on making spatial maps of hydrology probability forecasts throughout the United States or Canada, like the parent meteorology forecast maps. Furthermore, this book neither endorses the available weather forecasts nor expresses more confidence in some than in others. Readers must make their own judgments of the importance and priority of multiple meteorology probability forecasts, though the book provides some guidance. The book then helps readers use meteorology forecasts of their own selection to make derivative forecasts. Finally, while the examples in this book deal with long-range meteorology forecasts over a week, a month, or a 3-month period, set anywhere over the next 12 months, the methodology applies equally well to, and can immediately be used for, short-term forecasts.

Material of this type will be useful to water resource engineers, forecasters, and decision makers *only* if concepts are simply presented yet accompanied by sufficient depth in theory and applied principles. The writer has found that any new methodology must be clear and straightforward to be useful, yet engineers will not use it unless they also are comfortable with its underpinnings. Most recently, the writer has been working on these issues in small workshops for his laboratory's *Advanced Hydrologic Prediction System*, which utilizes these methods. He has also wrestled with these issues over the last 3 decades in both the university classroom and numerous university short courses for engineers. He has found that the two objectives of clear presentation of methodology and complete detail of theory and application can be difficult to achieve simultaneously. Very often, the mathematics and application details that are necessary for understanding a new methodology can obscure the explanation of it and of its use. It is often advantageous to separate the methodology presentation from the theory and application details.

This book is an expansion of ideas found in the writer's recently published works, in his lecture and tutorial notes for the workshops, and in his current work. In keeping with the philosophy of separate (but equal) presentations of methodology and sufficient theoretical underpinnings, the writer has organized the book in several ways. First, the book is divided into two parts. Part I presents the practical use and basic understanding of the forecasting techniques developed therein. Part I should enable practitioners to use the techniques, embodied in the free software described in the appendix, to make forecasts of their own based on available forecasts of meteorology probabilities. Part II of the book presents extensions of the methodology, much of the theoretical development, and real-world examples. It should enable the curious to understand why the methodology works, to appreciate its limitations, and to establish familiarity with some of the trade-offs involved. It also illustrates applications for all who would use the methodology.

The second way the writer has organized the material is to place selected material in "sidebars" or boxed sections, which can be skipped at first reading. Such material comprises additional details, supporting mathematics, or discussions of some extensions, which can aid understanding but are not strictly essential for learning and understanding the methodology. These materials are placed where they are most relevant but are organized for omission without impeding the flow of the concepts presented in this book. Readers may, at their discretion, elect to include the material on first exposure or return to it later, if interested, after gleaning a basic understanding.

The methodology entails duties that are easily handled by companion software, which is freely available. Otherwise, these duties are cumbersome in detail and could impede obtaining experience with the numerous examples presented in this book. The methodology requires readers to prepare historical data files for their applications and to tediously input forecast meteorology probabilities each time they make their own forecasts. It also requires readers to extract reference quantile estimates from the historical data and to prioritize the meteorology forecasts (and the resulting probability equations) in order of importance to them. Furthermore, it requires them to choose or define an objective for selecting a solution, and to solve the system of equations simultaneously (often iteratively while eliminating lowest-priority equations). The free computer software is in the form of a graphical user interface and enables the reader to easily perform these duties in an understandable and intuitive manner. Appendix 2 describes the acquisition and use of this software, contains illustrative examples for using it, and provides detailed instruction for duplicating all of the application examples in this book. The writer suggests that the reader study Part I and then acquire the software and become familiar with it by perusing the section "Usage" in Appendix 2. The reader can then revisit Chapters 4 and 5 while working the first four exercises, which are found in the "Usage" section in Appendix 2 but are frequently cross-referenced in Chapters 4 and 5. Then the reader should continue with Part II, referring as needed to the "Exercises" section in Appendix 2 when encountering each application example.

PART I

PRACTICAL USAGE

Chapter 1

INTRODUCTION

Estimating the probabilities of selected hydrology events is necessary for the design and operation of many water resource projects. Outlooks or forecasts of hydrology event probabilities allow water resource managers to assess the risks associated with their decisions in the construction and operation phases. On the basis of their own acceptable risk levels, managers can then make appropriate decisions regarding water resources. Probabilistic outlooks also allow managers to appreciate and quantify the uncertainty associated with forecasting, and to understand the wide range of future possibilities. These uncertainties include both future meteorology uncertainty and hydrology modeling uncertainty.

Multiple meteorology probability outlooks are now available to the water resource engineer or hydrologist, making more information available on meteorology uncertainty for derivative outlooks. Managers need to translate these new forecasts into derivative outlooks, such as hydrology probability outlooks, and that is the subject of this book.

BACKGROUND

One of the important problems in hydrology is interpreting a past record of events in terms of probabilities of future occurrence. (Actually, this interpretation of the past as future probabilities is of interest in many other geophysical engineering settings as well. Likewise, the discussion and developments made here for hydrology can be more broadly applied to these other settings.) Natural hydrology phenomena are variable but many times amenable to probabilistic interpretation and estimation analysis. Such interpretation requires that samples taken from the historical record be representative, unbiased, and independent. (Later in this chapter, the reader shall see that the "representative" and "unbiased" requirements can be replaced with only the presumption of an adequate range of possibilities.) "Representativeness" encompasses accurate observations that measure the item being represented over its range of possibilities, ergodicity (so one can use pieces of the record as an "ensemble" of series in estimation; i.e., there are no trends), stationarity in time series (implied by ergodicity, so one may expect that the future will be similar to the past in a probabilistic sense), and uniformity of record. "Unbiasedness" implies that the sampling procedure (used to pick an event or time series piece from the historical record as a possibility for the future) is equally likely to pick one observation as another. That is, all events or record pieces are equally likely as future events or time series, respectively. Thus, the use of the entire historical record presumes that the events, in the record, or the pieces of the record (whichever are being used) are defined so they can be considered equally likely to occur again in the future. "Independence" refers to the influence of one observation on another. Two successive floods occurring very close may result in a high degree of dependence of the second on the first. Thus, events or record pieces should be defined so they can be considered independent. Temporal dependence can be minimized by defining long event interarrival times or record pieces. For example, annual maximum floods or rainfalls (interarrival time on the order of a year) are often taken as time-independent, as are 1-year record pieces. Spatial independence must also be assured when multiple application areas are to be considered simultaneously.

There are two principal methods for determining hydrology probability forecasts or other derivative probability forecasts: (1) direct estimation from past hydrology records, and (2) estimation using past meteorology records transformed with appropriate hydrology models. The first method presumes that meteorology and watershed or other hydrology characteristics remain unchanged in the future. Examples include simple probability distribution estimation (as in estimating annual maximum flood frequency), co-axial correlation, principal components analysis, and time series analysis and modeling (as in fitting an auto-regressive moving average, or ARMA, model to monthly flow data). The second method presumes that meteorology characteristics remain unchanged and that process models ("derivative" models) suitably represent watershed or other hydrology characteristics. Thus, future hydrology conditions (or present ones) can be used with past records. Examples of the second method include Kalman filtering, Bayesian Forecasting, and "operational hydrology," where derivative process models like unit hydrographs, sophisticated watershed rainfall/runoff models, or other models are used to transform pieces of the historical meteorology record. The National Weather Service's Extended Streamflow Prediction (ESP) approach is one example of the operational hydrology approach.

The first method (direct estimation of hydrology event frequencies) is simpler than the second and seemingly more reliable. It enables engineers to obtain estimates of probabilities of hydrology events without detailed knowledge of the application area hydrology. It is also theoretically satisfying because one is working with observations of real events and not abstract process models. However, a short hydrology record limits the accuracy of estimation and may be unacceptably biased (e.g., those records where only wet periods or extended droughts are represented). While the operational hydrology method does require more work and sufficiently detailed knowledge of the application area hydrology, it allows the consideration of observable initial or boundary conditions, the accommodation of changes in hydrology, and the use of the generally much longer meteorology record.

The operational hydrology approach (also called the *Monte Carlo* approach) considers historical meteorology as possibilities for the future by segmenting the historical record and using each segment with derivative models and current initial conditions to simulate a hydrology possibility for the future ("hydrology scenario"). Each segment of the historical record then has associated time series of meteorology and hydrology variables, representing a possible scenario for the future. The resulting set of possible future scenarios is used to infer probabilities and other parameters associated with both meteorology and hydrology through estimation. A variation on this approach uses time series models of the meteorology to generate the resulting set of possible future scenarios. It must be remembered that multiple derivative future scenarios result not only from multiple meteorology future possibilities (selected historical record segments or time series models used as derivative model inputs), but also from other types of uncertainty. These other types include multiple possibilities for initial (current) conditions (such as uncertainty in observed soil moisture, snow pack, water temperatures, and so forth), observation or measurement variations (uncertainty) in the historical record, derivative model possibilities (model uncertainty), model calibration (parameter estimation uncertainty), and other factors. The uncertainty inherent in a derivative forecast encompasses all of these factors; this book examines the uncertainty associated with multiple meteorology future possibilities (*meteorology uncertainty*).

As mentioned earlier, multiple long-lead meteorology probability outlooks (forecasts of meteorology probabilities) are now available to the water resource engineer or hydrologist. These outlooks are defined over different time scales and time periods at different lag times, and they forecast either meteorology event probabilities or only most-probable meteorology events. It is possible to make probabilistic hydrology outlooks that use these existing meteorology outlooks by modifying the operational hydrology methodology. It is more desirable to do so now than at any time past, since meteorology outlooks have been improving and are now better than their predecessors. With the advent of probabilistic meteorology outlooks, more information is available for those wishing to use these outlooks in making derivative outlooks of their own, such as hydrology outlooks.

Those groups of meteorology segments (from the historical record or from time series models) matching forecast meteorology events are given more weight than those not matching, to agree with event probabilities, in constructing the resulting set of possible future derivative scenarios and in using them to estimate derivative probabilities. The historical meteorology record is thus used in a nonrepresentative and biased manner to match meteorology forecasts; therefore, the requirements are removed for representative and unbiased records and methodology. Still, the methodology requires that the historical record contain an adequate range of independent possibilities.

STRUCTURE OF BOOK

Part I of this book presents the methodology of using probabilistic meteorology outlooks to make derivative outlooks. It concentrates on practical knowledge and use, without much theoretical development, which is reserved for Part II. Knowledge of probability and estimation is useful to the understanding of relevant methodology, and a brief layman's review is presented first as Chapter 2. Those comfortable with probability and statistical concepts may skip Chapter 2, although it serves as the reference for definitions used in this book. Chapter 3 prepares for the later use of probabilistic outlooks by first discussing deterministic outlooks, along with their advantages and disadvantages. Probabilistic outlooks overcome the disadvantages of deterministic outlooks and are also discussed in Chapter 3 along with various specific useful estimates. Chapter 3 also delineates the types of probabilistic outlooks into "event probabilities" and "most-probable events." Chapter 4 considers some of the probabilistic meteorology outlooks that are now available to the public. There, one will find estimates of probabilities, for specific values of meteorology variables, published as outlooks by various agencies. In Chapter 4 and other chapters, nonessential material is presented in *sidebars* to the main text (boxed material). The casual reader who is interested only in understanding and using the methodology can skip this material. The sidebars provide more detail or a rigorous derivation of the equations for those desiring it. Chapter 5 discusses the transformation of meteorology outlooks into derivative hydrology outlooks, proceeding from simple deterministic outlooks through complex probabilistic outlooks. It introduces the concept of relating derivative forecast probabilities to a simple meteorology event probability and presents two examples. The chapter concludes with step-by-step instructions for using probabilistic meteorology outlooks to make a derivative outlook.

The instructions at the end of Chapter 5 illuminate the procedure that must be followed and introduce the reader to tools in Appendix 2 that can be used in practice.

However, while enabling work in real-world applications, the instructions provide insufficient discussion for complete understanding of the majority of real-world problems. Part II of this book extends the methodology, develops accompanying theory, and presents real-world examples for illustration. Chapter 6 introduces the concept of matching published multiple meteorology event probabilities by setting up and solving a system of equations, representing the probabilistic meteorology outlooks. Since multiple solutions usually exist, Chapter 7 presents methodology for choosing between competing meteorology outlooks and optimization for selection among multiple solutions. Chapter 8 extends the concept to matching both multiple meteorology event probabilities *and* most-probable events. Chapter 9 further extends the methodology to matching of multiple meteorology outlooks that are defined over different spatial extents. Chapter 10 introduces alternative optimizations that allow one to specify extremity of outlooks as an objective in the transformation from meteorology outlooks to a derivative hydrology outlook. Chapter 11 discusses evaluation issues and provides brief examples as guides for the reader's own evaluations. Chapters 6 through 10 are replete with examples, drawn from real-world experience or constructed for illustration. Appendix 1 provides a complete reference for all symbols used in this book. Appendix 2 provides documentation for freely available software, in the form of a graphical users interface for *Windows*, that allows readers to use available forecasts of meteorology probabilities in making their own derivative outlooks. Appendix 2 also gives directions for acquiring the software. It also contains all of the examples referred to in Chapters 4 through 10, completely worked with the software. With the software and Appendix 2 the reader will be able to work all of the examples without performing the tedious data entries that are part of the work. Appendix 3 contains development of sufficient conditions for the two minimization methods presented in Chapters 7 and 8. Relevant references are included in one section after the appendices.

Chapter 2

PROBABILITY AND ESTIMATION

A PROBABILITY MEASURE

Probability is a measure of the likelihood associated with an event. Suppose for example that there are 1 million flower seeds in a box and exactly 250,000 of them produce red flowers while the other 750,000 produce white flowers. If a person were to express the likelihood associated with the type of a seed arbitrarily selected from the box, he or she could say that the *probability* is 0.25 that the seed will produce a red flower. That is, 25% of the seed population produce red flowers and 75% produce white flowers. This probability measure applies for *equally likely outcomes:* the seeds are assumed to be thoroughly mixed, so that each seed is as likely to be selected as any other, without knowledge of which kind of flower it produces. The following discussion applies for this kind of probability measure, which measures likelihood as the relative proportion or *relative frequency* of equally likely possibilities of a certain type within a population.

OUTCOMES AND EVENTS

Let Ω denote the set of all possible outcomes in an experiment or the set of all possibilities in a population; it is often called the *target space* or *total population*. The elements of Ω are denoted by ω, and a particular value of ω is called an *outcome* or a *possibility*. The selection of ω corresponds to an outcome of the experiment or the realization of one of the possibilities from the population. A subset of Ω is called an *event* and includes some of the elements in Ω. The event A is said to *occur* if the selected outcome ω is in the set A. A set is denoted by notation in the form $\{\omega |$ a proposition is true$\}$, which reads "the set of all ω such that a proposition is true for each." Examples of statements of this form are "A = $\{\omega \mid \omega$ is in A$\}$" and "A = $\{\omega \mid \omega$ is not in B$\}$." Membership in a set is denoted in the form "$\omega \in$ A ," which is read "ω is in A." Combinations of events (sets) are themselves events (sets) and are denoted as follows:

$$A \cup B = \{\omega | \omega \in A \text{ or } \omega \in B\} \qquad \text{Union of sets} \qquad (2\text{-}1a)$$

$$A \cap B = AB = \{\omega | \omega \in A \text{ and } \omega \in B\} \quad \text{Intersection of sets} \qquad (2\text{-}1b)$$

$$\text{not A} = A^C = \{\omega | \omega \notin A\} \qquad \text{Complement of a set} \qquad (2\text{-}1c)$$

$$\varnothing = \{\omega | \omega \notin \Omega\} \qquad \text{Empty set} \qquad (2\text{-}1d)$$

Venn diagrams are useful depictions of the target space and various events. Figure 2-1 shows examples of the definitions in Equations (2-1). Inclusion of a set is denoted by "A $\subset$ B," which reads "A is contained in B,"

$$A \subset B \iff \omega \in A \implies \omega \in B \;\forall\; \omega \in A \qquad (2\text{-}2)$$

7

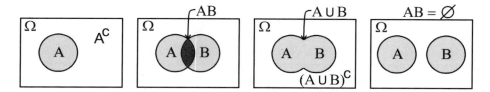

Figure 2-1. Example VENN diagrams.

where ⇔ denotes "if and only if," ⇒ denotes "implies," and ∀ denotes "for all." Equation (2-2) reads "A is contained in B if and only if an element's being in A implies it also is in B for all elements in A." Finally, set equality is defined as

$$A = B \iff A \subset B \text{ and } B \subset A \qquad (2\text{-}3)$$

The sets A and B are defined as disjoint or mutually exclusive if and only if $AB = \varnothing$; see Figure 2-1. The exclusive union of two sets A and B, denoted by "A ⊎ B," equals $A \cup B$ with the implication that A and B are mutually exclusive.

The probability of event A, $P[A]$, quantifies the likelihood of event A. As already mentioned, it is defined here as the fraction of the target space comprising event A, where the target space contains equally likely possibilities. The following intuitive axioms can be proven, although that is not done here; see, for example, Pfeiffer (1965):

$$P[A \uplus B \uplus C \uplus \cdots] = P[A] + P[B] + P[C] + \cdots \qquad (2\text{-}4a)$$

$$P[\varnothing] = 0 \qquad (2\text{-}4b)$$

$$P\left[A^c\right] = 1 - P[A] \qquad (2\text{-}4c)$$

$$0 \le P[A] \le 1 \qquad (2\text{-}4d)$$

$$P[A \cup B] = P[A] + P[B] - P[AB] \qquad (2\text{-}4e)$$

$$A \subset B \implies P[A] \le P[B] \qquad (2\text{-}4f)$$

Keeping in mind the definition of probability as a fraction of the target space, Equations (2-4) can be ascertained by inspection of Figure 2-1.

The *conditional probability* of A given B is defined as

$$P[A|B] = P[AB]/P[B] \qquad P[B] \ne 0 \qquad (2\text{-}5)$$

Conditional probability is interpreted as the probability of event A *given* that event B has occurred; it is seen intuitively to be the fraction of the set B occupied by set AB.

Events are considered *independent* of each other when one reveals nothing about the other. Alternatively,

$$P[A|B] = P[A] \qquad \Leftrightarrow \qquad \text{A and B are independent} \qquad (2\text{-}6)$$

or

$$P[AB] = P[A]P[B] \qquad \Leftrightarrow \qquad \text{A and B are independent} \qquad (2\text{-}7)$$

The theorem of total probability is

$$P[A] = P[A|B]P[B] + P[A|B^c]P[B^c] \qquad (2\text{-}8)$$

RANDOM VARIABLES AND DISTRIBUTIONS

Recall that a set is denoted in terms of a proposition, $\{\omega \mid$ a proposition is true$\}$. Consider now that the proposition is defined in terms of some function X that assigns a real number to every ω, $X(\omega)$. Thus, every possibility ω maps to a point on the real number line (has a corresponding value). This mapping function is referred to as a *random variable* and translates (assigns) every ω in Ω to the real number line. Events can then be expressed as, for example, $\{\omega \mid X(\omega) = 2\}$ or $\{\omega \mid X(\omega) \leq 3\}$ or, more generally, $\{\omega \mid X(\omega) \leq x\}$. The latter event employs a convention common in probability and estimation writings: uppercase denotes a function (called a random variable) and lowercase denotes a specific value of the function on the real number line. The probability of the latter event is $P\big[\{\omega \mid X(\omega) \leq x\}\big]$ and is denoted in a shorter expression as

$$P[X \leq x] = P\big[\{\omega \mid X(\omega) \leq x\}\big] \qquad (2\text{-}9)$$

This event, $X \leq x$ or $\{\omega \mid X(\omega) \leq x\}$, is of particular interest in probability theory. The first term in Equation (2-9) is referred to as the *cumulative distribution function* and has the following characteristics:

$$P[X \leq x_1] \leq P[X \leq x_2] \iff x_1 \leq x_2 \qquad (2\text{-}10a)$$
$$P[X \leq -\infty] = 0 \quad \text{or} \quad P[X \leq x] \geq 0 \qquad (2\text{-}10b)$$
$$P[X \leq \infty] = 1 \qquad (2\text{-}10c)$$

Equation (2-10a) says that the probability of $X \leq x_1$ (or $-\infty < X \leq x_1$) is smaller than or equal to the probability of $X \leq x_2$ (or $-\infty < X \leq x_2$) when the first interval, $(-\infty, x_1]$, is contained within the second, $(-\infty, x_2]$. This is similar to Equation (2-4f). Stated another way, Equation (2-10a) defines the cumulative distribution function to be monotonically non-decreasing in x and, together with the earlier observations, implies

$$P[x_1 < X \leq x_2] = P[X \leq x_2] - P[X \leq x_1] \qquad (2\text{-}11)$$

With the definition of Equation (2-9), Equation (2-7) implies that two random variables, X and Z, are independent if and only if

$$P[X \leq x \quad \text{and} \quad Z \leq z] = P[X \leq x]P[Z \leq z] \qquad (2\text{-}12)$$

where Z denotes another, different, mapping function. The left side of Equation (2-12) is referred to as the *joint* cumulative distribution function for random variables X and Z.

For a *continuous* random variable X, the derivative of the cumulative distribution function is called the *probability density function* $f_X(x)$, and the cumulative distribution function can be written as the integral of the probability density function:

$$f_X(x) = \frac{d}{dx}P[X \leq x] \qquad (2\text{-}13a)$$

$$P[X \leq x] = \int_{-\infty}^{x} f_X(u)\, du \tag{2-13b}$$

It can be shown that $P[X = x] = 0$ for any (and all) x for a continuous random variable X. This is because point x is one of an infinity of values on the real number line, and it has zero width, with no chance of being exactly realized. This is different from the *discrete* case. An example of a continuous random variable is air temperature. That is, there are values of air temperature on the real number line for all (of the infinite) possibilities ω within Ω.

For a discrete random variable X, there are probability masses (non-zero probabilities corresponding to specific values x). These are associated with each value of the random variable, and the cumulative distribution function can be written as

$$P[X \leq x] = \sum_{u=-\infty}^{x} P[X = u] \tag{2-14}$$

An example of a discrete random variable is a toss of the dice: there are only 12 possible values on the real line (1 through 12) and each possibility ω within Ω maps to one of these values. There are also mixed continuous and discrete random variables. One example is total precipitation, which is considered continuous for values greater than zero, but which has a probability mass associated with the value of zero (its lower bound).

DISTRIBUTION CHARACTERISTICS

The *expected value* of a function $g(X)$ of a random variable X is defined as the weighted integral or sum of all possible values (weighted by the distribution of X):

$$E[g(X)] = \int g(u) f_X(u)\, du \qquad \text{for continuous } X \tag{2-15a}$$

$$E[g(X)] = \sum g(u) P[X = u] \qquad \text{for discrete } X \tag{2-15b}$$

$$E[g(X)] = \int g(u) dP[X \leq u] \qquad \text{for mixed } X \tag{2-15c}$$

where the integrals or sum in Equations (2-15) are taken over all values of X and where $dP[X \leq u]$ is the change in the cumulative distribution function $P[X \leq x]$ as x goes to (approaches) u. For continuous portions of the distribution, $dP[X \leq u] = f_X(u)\, du$; and for discrete values, $dP[X \leq u] = P[X = u]$, and the integral becomes a sum at that point. Independence in Equation (2-12) can be written in terms of expected values. If random variables X and Z are independent, then

$$E[XZ] = E[X]E[Z] \tag{2-16}$$

The population *mean* μ_X, *variance* σ_X^2, *bias* BIAS_{XZ}, *mean square error* MSE_{XZ}, and *correlation* ρ_{XZ}, are defined in terms of expected values for random variables X and Z:

$$\mu_X = E[X] \tag{2-17a}$$

$$\sigma_X^2 = E\left[(X - \mu_X)^2\right] \tag{2-17b}$$

$$\text{BIAS}_{XZ} = E[X - Z] = \mu_X - \mu_Z \tag{2-17c}$$

$$MSE_{XZ} \;\; = \;\; E\left[(X - Z)^2\right] \tag{2-17d}$$

$$\rho_{XZ} \;\; = \;\; E\left[(X - \mu_X)(Z - \mu_Z)\right]\Big/\sqrt{\left(\sigma_X^2 \, \sigma_Z^2\right)} \tag{2-17e}$$

These are several of many population characteristics describing random variables or a pair of random variables. The mean is a measure of the center of the distribution of a random variable's values. The variance is a measure of the *spread* of the distribution about the mean. The bias measures the difference between the centers of the distributions of two different random variables. The mean square error reflects the discrepancy between two random variables' values, and the correlation is a measure of how well two random variables "track" each other in their joint occurrence. Unit correlation occurs when one random variable is a linear function of the other, and zero correlation occurs when one is completely (linearly) unrelated to the other. By Equation (2-16), if random variables X and Z are independent, then $\rho_{XZ} = 0$.

Another quantity related to cumulative distribution probabilities is the γ-probability quantile ξ_γ; it is defined as the value of the random variable for which the cumulative distribution function equals γ:

$$P\left[X \le \xi_\gamma\right] \;\; = \;\; \gamma \tag{2-18}$$

For example, if $P\left[X \le 12\right] \;\; = \;\; 0.1$, then the one-tenth quantile, $\xi_{0.1}$, is equal to 12. Since the cumulative distribution function is generally unknown in practice, parameter values in Equations (2-17) and (2-18) are unknown as well.

ESTIMATION

Information about a population or an experiment can be inferred in two manners. *Deductive* inference proceeds from knowledge of event membership. If outcome ω is in event A ($\omega \in$ A) and event A is contained in event B (A $\subset$ B), then outcome ω must be in event B too ($\omega \in$ B). For example, suppose (1) all West Point graduates (A) are over 18 years of age (B) and (2) John (ω) is a West Point graduate ($\omega \in$ A); since A $\subset$ B, then (3) John is deduced to be over 18 years of age ($\omega \in$ B). Results of deductive inference are definite. They are used to prove theorems, for example.

Inductive inference is a method of generalizing from a particular experiment to the class of all similar experiments. Consider again the example of 1 million flower seeds that was used earlier to introduce probability as a measure of likelihood. But now suppose that the numbers of seeds producing red or white flowers are unknown. It is infeasible and undesirable to plant all 1 million seeds to find the likelihood of red flowers. Instead a few seeds (a *sample*) are selected and planted and the number of red flower-producing seeds observed. This fraction is then generalized to the entire population. This is not an exact answer, because the sample may be (and usually is) an imperfect representation of the population, but it is an *estimate* that one may have confidence in. Conclusions reached by inductive inference are only *probable*, not definite. They are used to find new knowledge by observation. Inductive inference, or estimation through sampling, is used here to infer population parameters and to make probabilistic outlooks.

As already noted, in general the entire target space or population cannot be examined, but a *sample* of it can be, and inferences can be made from the sample. However, the *sample* must be selected in a certain fashion. Let $X_1, X_2, \dots, X_n$ represent successive experiments or observations about the target space or population, taken independently of each other. They constitute the sample. Suppose these variables have a joint cumulative distribution function that factors as follows:

$$P[X_1 \leq x_1, \ldots, X_n \leq x_n] = P[X_1 \leq x_1] \cdots P[X_n \leq x_n] = \prod_{i=1}^{n} F(x_i) \quad (2\text{-}19)$$

where $F(x)$ denotes the common (same) cumulative distribution function $P[X \leq x]$ for each variable. Thus Equation (2-19) identifies random variables that are independent and identically distributed (equally likely). Then $(X_1, X_2, \ldots, X_n)$ is said to be a *random sample* of size n from a population with a cumulative distribution function F. In practical parlance, the observations $(x_1, x_2, \ldots, x_n)$ are sometimes also called a random sample, where $x_1, x_2, \ldots, x_n$ are the values observed for $X_1, X_2, \ldots, X_n$, respectively. Functions of a random sample may be used as estimators (inductive inferences) of attributes of a target population. Discussion of the methods for deriving these estimators is beyond the scope of this text, but several estimators are of interest here for inferring the population parameters in Equations (2-17). They are the *sample mean* $\bar{x}$, *sample variance* s_X^2, *sample bias between X and Z* $\widehat{BIAS}_{XZ}$, *sample mean square error between X and Z* $\widehat{MSE}_{XZ}$, and *sample correlation between X and Z* $\hat{\rho}_{XZ}$:

$$\bar{x} = \frac{1}{n} \sum_{i=1}^{n} x_i \qquad\qquad (2\text{-}20a)$$

$$s_X^2 = \frac{1}{n} \sum_{i=1}^{n} (x_i - \bar{x})^2 \qquad\qquad (2\text{-}20b)$$

$$\widehat{BIAS}_{XZ} = \frac{1}{n} \sum_{i=1}^{n} (x_i - z_i) = \bar{x} - \bar{z} \qquad\qquad (2\text{-}20c)$$

$$\widehat{MSE}_{XZ} = \frac{1}{n} \sum_{i=1}^{n} (x_i - z_i)^2 \qquad\qquad (2\text{-}20d)$$

$$\hat{\rho}_{XZ} = \frac{\sum_{i=1}^{n} (x_i - \bar{x})(z_i - \bar{z})}{\sqrt{\sum_{i=1}^{n} (x_i - \bar{x})^2 \sum_{i=1}^{n} (z_i - \bar{z})^2}} \qquad\qquad (2\text{-}20e)$$

Equation (2-20e) uses a convention, common in probability and estimation writings, in which a caret ($^\wedge$) denotes an estimate of the population characteristic named under the caret. Thus, $\hat{\rho}_{XZ}$ is an estimate of ρ_{XZ}. There are other versions of these estimators, with slightly different properties; for example, an alternative form for the sample variance sometimes used in various applications is

$$s_X^2 = \frac{1}{n-1} \sum_{i=1}^{n} (x_i - \bar{x})^2 \qquad\qquad (2\text{-}21)$$

While Equations (2-20) are used in subsequent developments in this book, variations such as that in Equation (2-21) could also be used easily. All of these estimators are referred to as *nonparametric* because knowledge of the underlying distribution F and its parameters is not required. Other estimators (called *parametric*) derive from knowledge (or supposition) of the type of underlying distribution. Also useful is a nonparametric direct estimation of cumulative distribution probabilities. For example, the probability $P[X \leq x]$ is inferred with the estimator $\hat{P}[X \leq x]$, defined as the number of observations in the random sample, $(x_1, \ldots, x_n)$, that are less than or equal to x, $n_{X \leq x}$, divided by the total number of observations in the sample, n:

$$\hat{P}[X \leq x] = \frac{n_{X \leq x}}{n} \tag{2-22}$$

Alternatively,

$$\hat{P}[X \leq x] = \sum_A \frac{1}{n} \qquad A = \{i \mid x_i \leq x\} \tag{2-23a}$$

$$\hat{P}[X \leq x] = \frac{1}{n} \sum_{i \mid x_i \leq x} 1 \tag{2-23b}$$

In Equations (2-23), the count (sum) is taken over all i such that x_i is less than or equal to x. The estimator in Equations (2-23) is recognized as the *relative frequency* of $(x_i \leq x)$ in the random sample.

QUANTILE ESTIMATION
Suppose all values, x_i, in the random sample (x_i, $i = 1, \ldots, n$) are now ordered from smallest to largest to define the ordered variable values (y_m, $m = 1, \ldots, n$), where $y_m = x_{i(m)}$ and $i(m)$ is the number of the value in the unordered sample corresponding to the mth order (mth member of the ordered set). [For example, if the third value in the sample, x_3, is the largest ($y_n = x_3$), then $i(n) = 3$]. Then, Equations (2-23) give

$$\hat{P}[X \leq y_m] = \frac{m}{n} \qquad m = 1, \ldots, n \tag{2-24}$$

For example, if the fourth smallest value, y_4, is equal to 12 in a sample of size 40, then the estimated probability of not exceeding 12 is 4/40 = 0.1 (there are four values less than or equal to 12). The ordered value y_m is an estimate of $\xi_{m/n}$, defined in Equation (2-18). Continuing the example, $y_4 = 12$ is an estimate of the one-tenth quantile, $\xi_{0.1}$. Rewriting Equation (2-24) directly in terms of the random sample values,

$$\hat{P}\left[X \leq x_{i(m)}\right] = \frac{1}{n} \sum_{k=1}^{m} 1 \qquad m = 1, \ldots, n \tag{2-25}$$

Variations of Equation (2-24) are often used in hydrology to estimate points on cumulative distribution functions for annual exceedance series, such as annual maximum river flow. Some others are expressed in Chow (1964) in terms of exceedance probabilities ($P[X \geq y]$) on a sample ordered from largest to smallest. Rewriting for cumulative probability estimates on samples ordered from smallest to largest, they are:

$$\hat{P}[X \leq y_m] = \frac{m}{n} \qquad m = 1, \ldots, n, \qquad \text{California method} \tag{2-26a}$$

13

$$\hat{P}[X \le y_m] = \frac{m + \frac{1}{2}}{n} \qquad m = 1, \ldots, n, \qquad \text{Hazen method} \qquad (2\text{-}26b)$$

$$\hat{P}[X \le y_m] = \frac{m + 1}{n + 1} \qquad m = 1, \ldots, n, \qquad \text{Weibull method} \qquad (2\text{-}26c)$$

$$\hat{P}[X \le y_m] = \frac{m + 0.7}{n + 0.4} \qquad m = 1, \ldots, n, \qquad \text{Chegodayev method} \qquad (2\text{-}26d)$$

$$\hat{P}[X \le y_m] = \frac{m + \frac{5}{8}}{n + \frac{1}{4}} \qquad m = 1, \ldots, n, \qquad \text{Blom method} \qquad (2\text{-}26e)$$

$$\hat{P}[X \le y_m] = \frac{m + \frac{2}{3}}{n + \frac{1}{3}} \qquad m = 1, \ldots, n, \qquad \text{Tukey method} \qquad (2\text{-}26f)$$

$$\hat{P}[X \le y_m] = \frac{m + 0.56}{n + 0.12} \qquad m = 1, \ldots, n, \qquad \text{Gringorten method} \qquad (2\text{-}26g)$$

These estimators all have slightly different properties. While Equation (2-24) is used in subsequent developments in this book, any of the expressions in Equations (2-26) could also be easily used.

With the definitions and estimators presented in this chapter, various available useful estimates can now be discussed. Most particularly, estimates of cumulative distribution probabilities, for specific values of meteorology variables and for specific quantiles, are published as outlooks.

Chapter 3

OUTLOOK DEFINITIONS

DETERMINISTIC OUTLOOKS

Deterministic outlooks are forecasts of the value, or values, of a variable, or variables, expected in the future. They may be made, for example, by simply using historical averages or by using sophisticated models of pertinent physical processes applied to current conditions. Example deterministic meteorology outlooks include the following, made on September 1, 1998, for the month of September: "The average air temperature for the month in the area will be 15°C"; "the total precipitation for the month in the area will be 6 cm." These two examples can be expressed simply as:

$$T_{Sep98} = 15°C \qquad (3\text{-}1a)$$

$$Q_{Sep98} = 6 \ cm \qquad (3\text{-}1b)$$

where T_{Sep98} is average air temperature and Q_{Sep98} is total precipitation, both for September 1998. Example deterministic hydrology outlooks include the following, made on September 1, 1998, for the month of September: "The average basin moisture for the month in the watershed will be 21 cm"; "the total basin runoff for the month from the watershed will be 3 cm"; "the average water surface temperature for the month on the lake will be 4°C." These three examples also can be expressed simply as:

$$M_{Sep98} = 21 \ cm \qquad (3\text{-}2a)$$

$$R_{Sep98} = 3 \ cm \qquad (3\text{-}2b)$$

$$W_{Sep98} = 4°C \qquad (3\text{-}2c)$$

where M_{Sep98} is average basin moisture, R_{Sep98} is total basin runoff, and W_{Sep98} is average water surface temperature, all for September 1998. All of these examples are single-valued forecasts for single variables. One can also imagine single-valued forecasts for several variables at once. The preceding five single-valued single-variable examples are an example of a single-valued multiple-variable forecast when they are issued together.

There are also multiple-valued forecasts of a variable. For example, Figure 3-1 portrays a spatially varied multiple-valued deterministic meteorology outlook, a precipitation forecast over the United States made January 22, 1999, for the next 5 days. This could be used to generate a derivative hydrology forecast by transforming values at any point on the map with an appropriate rainfall runoff model, for example. Specific transformation examples are presented in Chapters 5 through 10. Figure 3-2 portrays another example: a temporally varied multiple-valued deterministic hydrology outlook. It was made September 14, 1998, for average monthly total moisture stored on the Lake Ontario watershed for the next 7 months (inclusive). One can also imagine multiple-valued multiple-variable forecasts by picturing spatial forecast maps, or forecasts of time series, for several variables issued together.

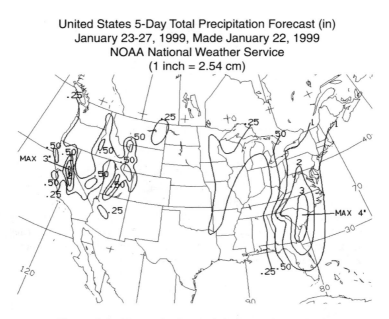

United States 5-Day Total Precipitation Forecast (in)
January 23-27, 1999, Made January 22, 1999
NOAA National Weather Service
(1 inch = 2.54 cm)

Figure 3-1. Example deterministic spatial outlook.

How these deterministic outlooks are generated is not really of concern here, but the method may range from simple extrapolation of recent trends to complex modeling and analysis of available data (specific examples are given in Chapter 5). Advantages of deterministic outlooks are that they are easy to understand, (relatively) easy to make, and easy to assess or evaluate as to "correctness" by observing what actually occurs. Disadvantages are they are difficult to make correctly, depending upon one's definition of correctness and they have low information content, in that they do not show a likelihood of

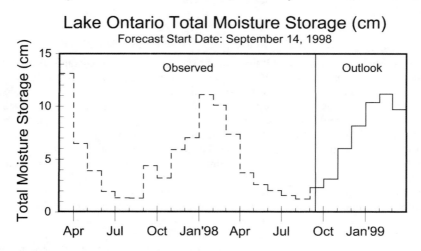

Figure 3-2. Example multiple-valued deterministic outlook for a single variable.

being close. Sometimes, several deterministic outlooks are made for a single variable to see what kind of "range" might be expected, but then there is the problem of selecting the "best." The probabilistic outlook overcomes these disadvantages.

PROBABILISTIC OUTLOOKS

Probabilistic outlooks are forecasts of a value, or values, of a variable, or variables, with each value or set of values associated with an estimated probability of some kind, reflecting the anticipated likelihood of an event defined in terms of the value(s) (such as an exceedance event). In other words, the forecast variables are considered to be random variables as defined in Chapter 2. An example of a probabilistic meteorology outlook made on September 1, 1998, for the month of September is, "The probability that the average air temperature for the month will exceed 15°C is estimated at 25%." This can be expressed simply as:

$$\hat{P}\left[T_{\text{Sep98}} > 15°C\right] = 0.25 \tag{3-3}$$

Other examples, also made on September 1 for the month of September, include the following: "The probability that the total precipitation for the month will be below 5 cm is estimated at 40%"; "the probability that the total runoff for the month will be less than or equal to 3.5 cm is estimated at 65%"; "the probability that the average water surface temperature will be between 3.4°C and 4.2°C is estimated at 90%." These can be expressed, respectively, as:

$$\hat{P}\left[Q_{\text{Sep98}} < 5 \text{ cm}\right] = 0.40 \tag{3-4a}$$

$$\hat{P}\left[R_{\text{Sep98}} \leq 3.5 \text{ cm}\right] = 0.65 \tag{3-4b}$$

$$\hat{P}\left[3.4°C < W_{\text{Sep98}} < 4.2°C\right] = 0.90 \tag{3-4c}$$

These examples express the likelihood of forecast events in terms of *event probabilities*, that is, the probabilities that the forecast events will occur. An alternative probabilistic outlook type expresses likelihood in terms of *most-probable events*. In this case, an event is presented with an indication of its anticipation as the most likely event. An example of a probabilistic meteorology outlook of this type is, "An average (over the month) air temperature higher than the median (50% quantile) is expected for September 1998." Interpreting this statement allows this alternative expression: "The average air temperature for September will be greater than the (estimated historical) median with a probability greater than that usually associated with the median." By the definition in Equation (2-18), there is a probability of 50% of not exceeding the median, $\tau_{\text{Sep}, 0.50}$.

The median may then be estimated $\left(\hat{\tau}_{\text{Sep}, 0.50}\right)$ as that value not exceeded 50% of the time in the historical record:

$$\hat{P}\left[T_{\text{Sep}} \leq \hat{\tau}_{\text{Sep}, 0.50}\right] = 0.50 \tag{3-5}$$

where T_{Sep} is average September air temperature. That means there is a 50% chance in the historical record associated with exceeding the estimated median:

$$\hat{P}\left[T_{\text{Sep}} > \hat{\tau}_{\text{Sep}, 0.50}\right] = 1 - \hat{P}\left[T_{\text{Sep}} \leq \hat{\tau}_{\text{Sep}, 0.50}\right] = 0.50 \tag{3-6}$$

Thus, the example probabilistic temperature outlook for September 1998 can be expressed as:

$$\hat{P}\left[T_{\text{Sep98}} > \hat{\tau}_{\text{Sep},0.50}\right] > 0.50 \tag{3-7}$$

Likewise, the probabilistic outlook, "a total precipitation below the median is expected for September 1998," could be interpreted as follows:

$$\hat{P}\left[Q_{\text{Sep98}} \leq \hat{\theta}_{\text{Sep},0.50}\right] > 0.50 \tag{3-8}$$

where $\hat{\theta}_{\text{Sep},0.50}$ is defined in a way similar to $\hat{\tau}_{\text{Sep},0.50}$ as the estimated median September total precipitation, calculated from the historical record. [In general, $\hat{\theta}_{\text{Sep},\gamma}$ is defined in a form similar to Equation (2-18), as the reference estimated precipitation γ-probability quantile for September that satisfies:

$$\hat{P}\left[Q_{\text{Sep}} \leq \hat{\theta}_{\text{Sep},\gamma}\right] = \gamma \tag{3-9}$$

where Q_{Sep} is total September precipitation.]

Single-valued multiple-variable probabilistic outlooks can also be achieved by issuing several single-variable outlooks simultaneously, in a way similar to the deterministic case. There are also multiple-valued probabilistic forecasts of a variable. A common example is the spatial probability map. Figure 3-3 portrays an example, for the estimated probability that September 1998 total precipitation will be within the lower third of its historical range. For example, at the asterisk near the Mexican border in Figure 3-3, the following is forecast:

$$\hat{P}\left[Q_{\text{Sep98}} \leq \hat{\theta}_{\text{Sep},0.333}\right] = 0.43 \tag{3-10}$$

where $\hat{\theta}_{\text{Sep},0.333}$ is the estimated one-third quantile for September total precipitation, defined in Equation (3-9) for $\gamma = 0.333$.

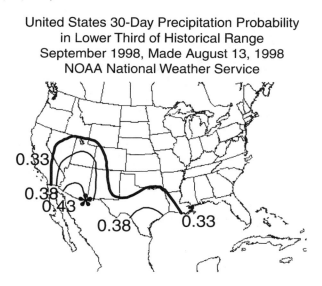

United States 30-Day Precipitation Probability
in Lower Third of Historical Range
September 1998, Made August 13, 1998
NOAA National Weather Service

Figure 3-3. Example probabilistic spatial outlook (event probabilities).

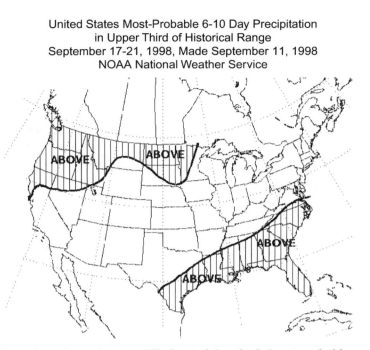

United States Most-Probable 6-10 Day Precipitation
in Upper Third of Historical Range
September 17-21, 1998, Made September 11, 1998
NOAA National Weather Service

Figure 3-4. Example probabilistic spatial outlook (most probable event).

While Figure 3-3 portrays a spatial map example defined in terms of event probabilities, Figure 3-4 illustrates a spatial probabilistic forecast defined in terms of a most-probable event. It shows where, in the United States, 6–10 day precipitation greater than the upper third of its historical range is forecast as most probable. Interpreting Figure 3-4 in the manner of Equation (3-7) for the state of Florida, but for the estimated two-thirds quantile $\hat{\theta}_{17-21\text{Sep},\,0.667}$ instead of the estimated median,

$$\hat{P}\left[Q_{17-21\text{Sep}98} > \hat{\theta}_{17-21\text{Sep},\,0.667}\right] > 0.333 \tag{3-11}$$

There are also forecasts of entire distributions for a variable, rather than simply probabilities associated with a single event or a most-probable event. A typical example also uses multiple-valued forecasts of a variable that are temporally varying. Figure 3-5 shows an example multiple-valued probabilistic hydrology outlook made September 14, 1998. It portrays estimated Lake Ontario basin moisture storage probability distributions for the next 6 or 7 months. This forecast estimates cumulative distribution values each month as a set of bands, one contained within another. The single line in the middle of the bands denotes the value forecast not to be exceeded with a probability of 0.50:

$$\hat{P}\left[M_g \leq m_g\right] = 0.50 \tag{3-12}$$

where g denotes the month of the forecast (September 1998 through March 1999). The inner-most band (30–70%) shows all values of storage such that:

$$0.30 \leq \hat{P}\left[M_g \leq m_g\right] \leq 0.70 \tag{3-13}$$

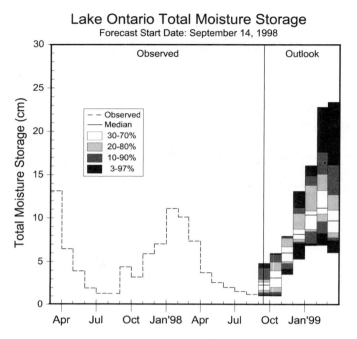

Figure 3-5. Example multiple-valued probabilistic outlook for a single variable.

Alternatively, the lower limit of this band (at 30%) and the upper limit of this band (at 70%) represent, respectively, the estimates

$$\hat{P}\left[M_g \le m_g \right] = 0.30 \tag{3-14a}$$

$$\hat{P}\left[M_g \le m_g \right] = 0.70 \tag{3-14b}$$

The other bands have a similar correspondence:

$$0.20 \le \hat{P}\left[M_g \le m_g \right] \le 0.80 \qquad (20-80\%) \tag{3-15a}$$

$$0.10 \le \hat{P}\left[M_g \le m_g \right] \le 0.90 \qquad (10-90\%) \tag{3-15b}$$

$$0.03 \le \hat{P}\left[M_g \le m_g \right] \le 0.97 \qquad (3-97\%) \tag{3-15c}$$

   In general, the disadvantages of probabilistic outlooks are that they are difficult to understand, difficult to make, and difficult to assess. However, they obviate the need to select a "best" deterministic outlook. More importantly, they have high information content, maximizing the use of available information and providing more information for decision-makers. For example, water resource managers can translate forecast probabilities of critical meteorology events into expressions of risks to associate with derivative critical events of their own. Based, then, on the managers' acceptable risk levels, decisions can be made on the operation of their water resource works. Managers first need a way to translate probabilistic meteorology outlooks into hydrology or other derivative outlooks. That is the subject of Chapter 5. But first, it is appropriate to consider some of the probabilistic meteorology outlooks that are available to the public.

Chapter 4

PROBABILISTIC METEOROLOGY OUTLOOK EXAMPLES

NATIONAL OCEANIC AND ATMOSPHERIC ADMINISTRATION (NOAA) CLIMATE OUTLOOKS

Multiple probabilistic meteorology outlooks can consist of multiple forms of probability outlooks: event probabilities and most-probable events. There are now several kinds of event probability outlooks available to the water resource engineer or hydrologist for making derivative forecasts. One of the earliest still available is the *Climate Outlook*, provided by the National Oceanic and Atmospheric Administration (NOAA) Climate Prediction Center (CPC), which is one of the NOAA National Centers for Environmental Prediction (NCEPs). The *Climate Outlook* is available over the World Wide Web via links from the NOAA home page to the CPC products page. The CPC provides this outlook each month at approximately mid-month; it consists of a 1-month outlook for the next (full) month and thirteen 3-month outlooks, going into the future in overlapping fashion in 1-month steps.

Background and recent history on seasonal forecasting are provided in Barnston et al. (1994), van den Dool (1994), Livezey (1990), Wagner (1989), Epstein (1988), Ropelewski and Halpert (1986), and Gilman (1985). The forecasts in the *Climate Outlook* are formed by a combination of methods. For US air temperature and precipitation forecasts, these methods included, as of 1994, (1) canonical correlation analysis (Barnston and Ropelewski 1992) relating spatial anomalies of sea surface temperature in selected regions, Northern Hemisphere 700-mb height, and the US surface climate (referred to as "teleconnections"); and also (2) observed interannual persistence of anomalies (Huang et al. 1994), as well as (3) forecasts from two 6-month atmospheric general circulation models driven by sea surface temperatures. One is a set continued from one-half month earlier and the other is a set assembled from coupled ocean-atmosphere model runs (Ji et al. 1994). The general circulation model is a version of the NCEP Environmental Modeling Center's (EMC's) Medium-Range Forecast Model, which has a global domain, with special developmental emphasis on tropical processes.

Each outlook estimates probabilities of average air temperature and total precipitation falling within preselected value ranges. The value ranges (low, normal, and high) are defined as the lower, middle, and upper thirds of observations over the period 1961–90 for each variable. The climate outlooks presume that one of only four possibilities exists for the probability distribution type for each variable: (1) the probability of being in the high range exceeds one-third, and the probability of being in the low range is reduced accordingly (it remains at one-third for the normal range)—referred to as being "above normal"; (2) the probability of being in the normal range exceeds one-third, and the probabilities of being in the low and high ranges are reduced accordingly and are equal—referred to as being "normal"; (3) the probability of being in the low range exceeds one-third, and the probability of being in the high range is reduced accordingly (it remains at one-third for the normal range)—referred to as being "below normal"; or (4) skill is insufficient to make a forecast, and so probabilities of one-third in each range are

21

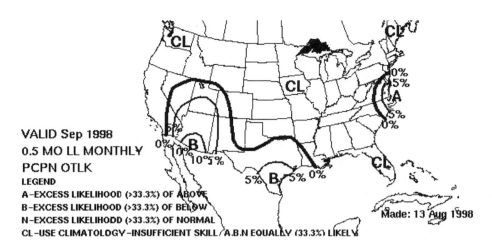

VALID Sep 1998
0.5 MO LL MONTHLY
PCPN OTLK
LEGEND
A-EXCESS LIKELIHOOD (>33.3%) OF ABOVE
B-EXCESS LIKELIHOOD (>33.3%) OF BELOW
N-EXCESS LIKELIHODD (>33.3%) OF NORMAL
CL-USE CLIMATOLOGY-INSUFFICIENT SKILL/A.B.N EQUALLY (33.3%) LIKELY

Made: 13 Aug 1998

Figure 4-1. NOAA 1-month probabilistic precipitation outlook for September 1998.

used—referred to as "climatological." This "four-distribution universe" is a built-in definition associated with using NOAA's *Climate Outlook* and, while not always a realistic model of natural distributions, must be used in interpreting their outlooks.

An example outlook of precipitation probabilities is shown in Figure 4-1, representing the NOAA 1-month climatic outlook for September 1998, made on August 13, 1998. It is shown in an alternative form in Figure 4-2, which depicts only the continental United States and has less map detail, clarifying the presentation somewhat. The actual *Climate Outlook* is presented in color with all probabilities mapped together. They are separated here, because gray scales for probabilities "above" and "below" are confusing if plotted together. Forecast probabilities can be ascertained for any point on the outlook map. For example, in mid-Texas in Figure 4-1 or 4-2, the probability of September precipitation in the lower third of historical observations is forecast to rise by about 0.02. According to the convention of NOAA's definitions, the corresponding probability of September precipitation in the upper third of historical observations is forecast to drop by about 0.02, and the probability of September precipitation in the middle third of historical observations is forecast to remain unchanged at one-third:

$$\hat{P}\left[Q_{Sep98} \leq \hat{\theta}_{Sep, 0.333}\right] = 0.333 + 0.02 = 0.353 \tag{4-1a}$$

$$\hat{P}\left[\hat{\theta}_{Sep, 0.333} < Q_{Sep98} \leq \hat{\theta}_{Sep, 0.667}\right] = 0.334 \tag{4-1b}$$

$$\hat{P}\left[Q_{Sep98} > \hat{\theta}_{Sep, 0.667}\right] = 0.333 - 0.02 = 0.313 \tag{4-1c}$$

where $\hat{\theta}_{Sep, \gamma}$ now denotes the reference γ-probability quantile estimate, calculated from the 1961–90 historical record for September with Equation (3-9).

Actually, there are multiple 1-month and 3-month outlooks every month from NOAA; Figure 4-2 is but one of 28 available each month, as shown in the full *Climate Outlook* depicted in Figures 4-3 and 4-4. (Again, the originals are presented in color; the probabilities for "above" and "below" are separated here, because gray scales for the two are confusing if plotted together.) These figures represent a large number of probability forecast statements. At any one particular site, the multiple outlooks consist of

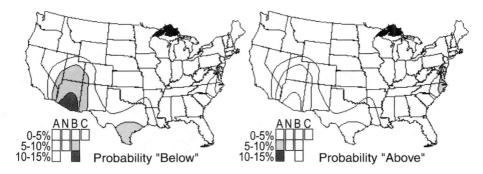

Figure 4-2. Alternative-form of NOAA 1-month probabilistic precipitation outlook for September 1998.

one 1-month and thirteen 3-month outlooks of both temperature and precipitation, each with three equations per outlook for each variable.

$$\hat{P}\left[T_g \leq \hat{\tau}_{g, 0.333} \right] = a_g \qquad g = 1, \dots, 14 \quad (4\text{-}2a)$$

$$\hat{P}\left[T_g > \hat{\tau}_{g, 0.667} \right] = b_g \qquad g = 1, \dots, 14 \quad (4\text{-}2b)$$

$$\hat{P}\left[\hat{\tau}_{g, 0.333} < T_g \leq \hat{\tau}_{g, 0.667} \right] = 1 - a_g - b_g \qquad g = 1, \dots, 14 \quad (4\text{-}2c)$$

$$\hat{P}\left[Q_g \leq \hat{\theta}_{g, 0.333} \right] = c_g \qquad g = 1, \dots, 14 \quad (4\text{-}2d)$$

$$\hat{P}\left[Q_g > \hat{\theta}_{g, 0.667} \right] = d_g \qquad g = 1, \dots, 14 \quad (4\text{-}2e)$$

$$\hat{P}\left[\hat{\theta}_{g, 0.333} < Q_g \leq \hat{\theta}_{g, 0.667} \right] = 1 - c_g - d_g \qquad g = 1, \dots, 14 \quad (4\text{-}2f)$$

where T_g and Q_g are average air temperature and total precipitation at the site, respectively, over period g ($g = 1$ corresponds to a 1-month period, and $g = 2, \dots, 14$ corresponds to 13 successive overlapping 3-month periods); $\hat{\tau}_{g,\gamma}$ and $\hat{\theta}_{g,\gamma}$ are, respectively, temperature and precipitation reference γ-probability quantile estimates at the site for period g defined similarly to Equation (3-9); and (a_g, b_g, c_g, and d_g, $g = 1, \dots, 14$) are the outlook probability forecast estimates as calculated from the map readings at the site. Recall that the reference γ-probability quantiles are estimated from the 1961–90 historical record at the site for each period g by definition. For the September 1998 *Climate Outlook* in Figures 4-3 and 4-4 at any site, there is a 1-month September outlook ($g = 1$ or "Sep") and thirteen 3-month outlooks successively lagged by 1 month each ($g = 2$ or "September-October-November" or "SON," and $g = 3, 4, \dots, 14$ or "OND," "NDJ," ... , "SON," respectively).

Equations (4-2c) and (4-2f) are redundant in combination with the rest of Equations (4-2) because probabilities (and probability estimates) sum to unity over the real line:

$$\hat{P}\left[T_g \leq \hat{\tau}_{g, 0.333} \right] + \hat{P}\left[\hat{\tau}_{g, 0.333} < T_g \leq \hat{\tau}_{g, 0.667} \right] + \hat{P}\left[T_g > \hat{\tau}_{g, 0.667} \right] = 1 \quad (4\text{-}3a)$$

$$\hat{P}\left[Q_g \leq \hat{\theta}_{g, 0.333} \right] + \hat{P}\left[\hat{\theta}_{g, 0.333} < Q_g \leq \hat{\theta}_{g, 0.667} \right] + \hat{P}\left[Q_g > \hat{\theta}_{g, 0.667} \right] = 1 \quad (4\text{-}3b)$$

Therefore there are four independent settings in Equations (4-2) at any site for each of the 14 climate outlooks, for a total of 56. Example numbers, taken from the September

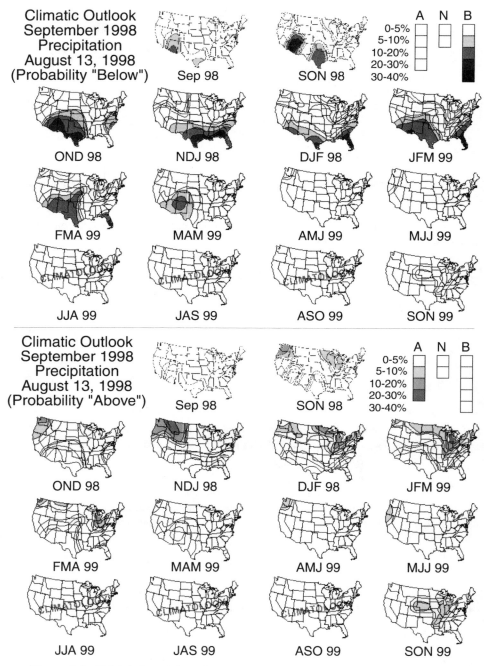

Figure 4-3. NOAA 1-month and extended 3-month probabilistic precipitation outlook for September 1998 and September-October-November 1998 through September-October-November 1999.

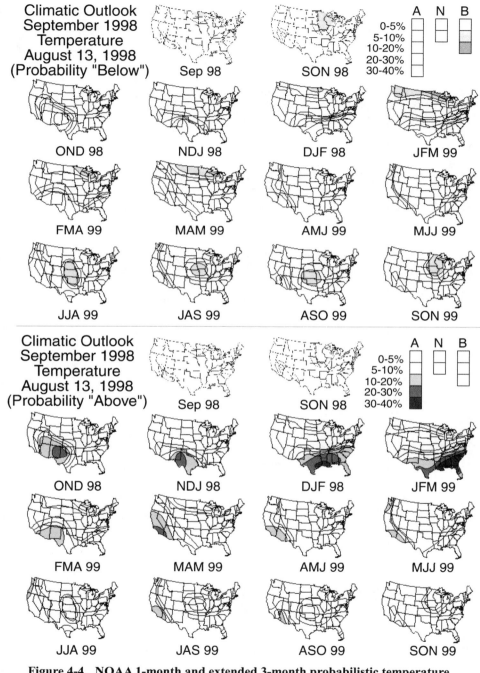

Figure 4-4. **NOAA 1-month and extended 3-month probabilistic temperature outlook for September 1998 and September-October-November 1998 through September-October-November 1999.**

25

Table 4-1. NOAA September 1998 Lake Superior basin outlook event probability estimates (%).

Period, g	$\hat{P}_T$ [a]	$\hat{P}_Q$ [a]	Temperature Probabilities [b]			Precipitation Probabilities [b]		
			$(-\infty, \hat{\tau}_{g,0.333}]$	$(\hat{\tau}_{g,0.333}, \hat{\tau}_{g,0.667}]$	$(\hat{\tau}_{g,0.667}, \infty)$	$(-\infty, \hat{\theta}_{g,0.333}]$	$(\hat{\theta}_{g,0.333}, \hat{\theta}_{g,0.667}]$	$(\hat{\theta}_{g,0.667}, \infty)$
(1)	(2)	(3)	(4)	(5)	(6)	(7)	(8)	(9)
Sep 98	1 b	0 c	34.3	33.4	32.3	33.3	33.4	33.3
SON 98	5 b	5 a	38.3	33.4	28.3	28.3	33.4	38.3
OND 98	0 c	3 a	33.3	33.4	33.3	30.3	33.4	36.3
NDJ 98	0 c	3 a	33.3	33.4	33.3	30.3	33.4	36.3
DJF 98	0 c	11 a	33.3	33.4	33.3	22.3	33.4	44.3
JFM 99	11 b	10 a	44.3	33.4	22.3	23.3	33.4	43.3
FMA 99	6 b	1 a	39.3	33.4	27.3	32.3	33.4	34.3
MAM 99	7 b	0 c	40.3	33.4	26.3	33.3	33.4	33.3
AMJ 99	0 c	0 c	33.3	33.4	33.3	33.3	33.4	33.3
MJJ 99	0 c	0 c	33.3	33.4	33.3	33.3	33.4	33.3
JJA 99	0 c	0 c	33.3	33.4	33.3	33.3	33.4	33.3
JAS 99	0 c	0 c	33.3	33.4	33.3	33.3	33.4	33.3
ASO 99	0 c	0 c	33.3	33.4	33.3	33.3	33.4	33.3
SON 99	6 b	0 c	39.3	33.4	27.3	33.3	33.4	33.3

[a] Probability estimates ($\hat{P}_T$ and $\hat{P}_Q$ designate temperature and precipitation probability estimates, respectively) in excess of 33.3% in low interval (below normal), in mid interval (normal), or in high interval (above normal); "no forecast" is indicated by "0 c" (climatological).

[b] Probability estimates over NOAA's CPC's corresponding interval definitions.

1998 maps for the Lake Superior basin, appear in columns 2 and 3 in Table 4-1. (Note that, in Figures 4-1 and 4-2, Lake Superior has been blackened to highlight its location. This is not repeated on all of the other forecast maps following, but this location will be used in many examples to follow.) They are interpreted as increments or decrements to appropriate reference values to yield probability estimates in columns 4 through 9. Note that column 5 is redundant in combination with columns 4 and 6, as is column 8 with columns 7 and 9. The probability estimate equations represented by Figures 4-3 and 4-4 and Table 4-1 are listed in Figure 4-5. See Exercises A2-1, A2-2, and A2-2a in Appendix 2. (Now is a good time to acquire the public domain software documented in Appendix 2 and to learn its use. It will make it possible to work all of the real-world examples in this book and eliminate the tedium of data entry for the examples. It will also make it possible to conveniently enter forecast meteorology probabilities and allow readers to later use them in creating their own derivative outlooks.)

NOAA 8–14 DAY EVENT PROBABILITY OUTLOOKS

The NOAA CPC also makes an experimental "second-week" outlook at about 3:00 p.m. EST every Thursday that gives probabilities for average air temperature and total precipitation events over a 7-day period beginning 7 days from the forecast date. Therefore, there are four independent settings in this outlook. NOAA calls this its *8–14 day outlook,* and it is based on teleconnections, persistence, and model simulations using the medium-range forecast model, similar to the generation of their *Climate Outlook.* Each 8–14 day outlook provides probability estimates of average air temperature and total precipitation falling within preselected value ranges of low, normal, and high, again defined as the lower, middle, and upper thirds of observations over the period 1961–90 for each variable. The climate outlooks again presume the "four-distribution universe" that was defined before. Recall that only four possibilities are assumed to exist for the probability distributions for each variable: (1) "above normal" (probability of high range exceeds one-third, with probability of low reduced accordingly), (2) "normal" (probability of middle range exceeds one-third, with probabilities of being low and high reduced accordingly and equally), (3) "below normal" (probability of low range exceeds one-third, with probability of high reduced accordingly), and (4) "climatological" (probabilities of one-third in each range are used).

An example is pictured in Figure 4-6 for September 11–17, 1998; it was forecast on September 3, 1998, and downloaded from NOAA's CPC Web site. Example numbers are taken from these maps for the Lake Superior basin: 0.18 above for air temperature and 0.06 below for precipitation. They are interpreted as increments or decrements to the appropriate reference values to yield probability estimate equations, similar to reading and using NOAA's 1- and 3-month outlooks. The resulting probability estimate equations are

$$\hat{P}\left[T_{11-17\text{Sep}98} \leq \hat{\tau}_{11-17\text{Sep},0.333}\right] \quad = \quad 0.333 - 0.18 \quad = \quad 0.153 \tag{4-4a}$$

$$\hat{P}\left[T_{11-17\text{Sep}98} > \hat{\tau}_{11-17\text{Sep},0.667}\right] \quad = \quad 0.333 + 0.18 \quad = \quad 0.513 \tag{4-4b}$$

$$\hat{P}\left[Q_{11-17\text{Sep}98} \leq \hat{\theta}_{11-17\text{Sep},0.333}\right] \quad = \quad 0.333 + 0.06 \quad = \quad 0.393 \tag{4-4c}$$

$$\hat{P}\left[Q_{11-17\text{Sep}98} > \hat{\theta}_{11-17\text{Sep},0.667}\right] \quad = \quad 0.333 - 0.06 \quad = \quad 0.273 \tag{4-4d}$$

See Exercises A2-1, A2-2, and A2-2b in Appendix 2.

Another 8–14 day outlook, available from NOAA's Environmental Research Laboratory's Climate Diagnostic Center (CDC), is the week 2 tercile precipitation probabili-

Lake Superior Basin Event Probability Estimates
September 1998 Air Temperature & Precipitation
SON 1998 through SON 1999 Air Temperature & Precipitation
Forecast August 13, 1998 by NOAA

$\hat{P}\left[T_{Sep98} \leq \hat{\tau}_{Sep,0.333}\right] = 0.343$

$\hat{P}\left[T_{Sep98} > \hat{\tau}_{Sep,0.667}\right] = 0.323$

$\hat{P}\left[Q_{Sep98} \leq \hat{\theta}_{Sep,0.333}\right] = 0.333$

$\hat{P}\left[Q_{Sep98} > \hat{\theta}_{Sep,0.667}\right] = 0.333$

$\hat{P}\left[T_{SON98} \leq \hat{\tau}_{SON,0.333}\right] = 0.383$

$\hat{P}\left[T_{SON98} > \hat{\tau}_{SON,0.667}\right] = 0.283$

$\hat{P}\left[Q_{SON98} \leq \hat{\theta}_{SON,0.333}\right] = 0.283$

$\hat{P}\left[Q_{SON98} > \hat{\theta}_{SON,0.667}\right] = 0.383$

$\hat{P}\left[T_{OND98} \leq \hat{\tau}_{OND,0.333}\right] = 0.333$

$\hat{P}\left[T_{OND98} > \hat{\tau}_{OND,0.667}\right] = 0.333$

$\hat{P}\left[Q_{OND98} \leq \hat{\theta}_{OND,0.333}\right] = 0.303$

$\hat{P}\left[Q_{OND98} > \hat{\theta}_{OND,0.667}\right] = 0.363$

$\hat{P}\left[T_{NDJ98} \leq \hat{\tau}_{NDJ,0.333}\right] = 0.333$

$\hat{P}\left[T_{NDJ98} > \hat{\tau}_{NDJ,0.667}\right] = 0.333$

$\hat{P}\left[Q_{NDJ98} \leq \hat{\theta}_{NDJ,0.333}\right] = 0.303$

$\hat{P}\left[Q_{NDJ98} > \hat{\theta}_{NDJ,0.667}\right] = 0.363$

$\hat{P}\left[T_{DJF98} \leq \hat{\tau}_{DJF,0.333}\right] = 0.333$

$\hat{P}\left[T_{DJF98} > \hat{\tau}_{DJF,0.667}\right] = 0.333$

$\hat{P}\left[Q_{DJF98} \leq \hat{\theta}_{DJF,0.333}\right] = 0.223$

$\hat{P}\left[Q_{DJF98} > \hat{\theta}_{DJF,0.667}\right] = 0.443$

$\hat{P}\left[T_{JFM99} \leq \hat{\tau}_{JFM,0.333}\right] = 0.443$

$\hat{P}\left[T_{JFM99} > \hat{\tau}_{JFM,0.667}\right] = 0.223$

$\hat{P}\left[Q_{JFM99} \leq \hat{\theta}_{JFM,0.333}\right] = 0.233$

$\hat{P}\left[Q_{JFM99} > \hat{\theta}_{JFM,0.667}\right] = 0.433$

$\hat{P}\left[T_{FMA99} \leq \hat{\tau}_{FMA,0.333}\right] = 0.393$

$\hat{P}\left[T_{FMA99} > \hat{\tau}_{FMA,0.667}\right] = 0.273$

$\hat{P}\left[Q_{FMA99} \leq \hat{\theta}_{FMA,0.333}\right] = 0.323$

$\hat{P}\left[Q_{FMA99} > \hat{\theta}_{FMA,0.667}\right] = 0.343$

$\hat{P}\left[T_{MAM99} \leq \hat{\tau}_{MAM,0.333}\right] = 0.403$

$\hat{P}\left[T_{MAM99} > \hat{\tau}_{MAM,0.667}\right] = 0.263$

$\hat{P}\left[Q_{MAM99} \leq \hat{\theta}_{MAM,0.333}\right] = 0.333$

$\hat{P}\left[Q_{MAM99} > \hat{\theta}_{MAM,0.667}\right] = 0.333$

$\hat{P}\left[T_{AMJ99} \leq \hat{\tau}_{AMJ,0.333}\right] = 0.333$

$\hat{P}\left[T_{AMJ99} > \hat{\tau}_{AMJ,0.667}\right] = 0.333$

$\hat{P}\left[Q_{AMJ99} \leq \hat{\theta}_{AMJ,0.333}\right] = 0.333$

$\hat{P}\left[Q_{AMJ99} > \hat{\theta}_{AMJ,0.667}\right] = 0.333$

$\hat{P}\left[T_{MJJ99} \leq \hat{\tau}_{MJJ,0.333}\right] = 0.333$

$\hat{P}\left[T_{MJJ99} > \hat{\tau}_{MJJ,0.667}\right] = 0.333$

$\hat{P}\left[Q_{MJJ99} \leq \hat{\theta}_{MJJ,0.333}\right] = 0.333$

$\hat{P}\left[Q_{MJJ99} > \hat{\theta}_{MJJ,0.667}\right] = 0.333$

$\hat{P}\left[T_{JJA99} \leq \hat{\tau}_{JJA,0.333}\right] = 0.333$

$\hat{P}\left[T_{JJA99} > \hat{\tau}_{JJA,0.667}\right] = 0.343$

$\hat{P}\left[Q_{JJA99} \leq \hat{\theta}_{JJA,0.333}\right] = 0.333$

$\hat{P}\left[Q_{JJA99} > \hat{\theta}_{JJA,0.667}\right] = 0.333$

$\hat{P}\left[T_{JAS99} \leq \hat{\tau}_{JAS,0.333}\right] = 0.333$

$\hat{P}\left[T_{JAS99} > \hat{\tau}_{JAS,0.667}\right] = 0.333$

$\hat{P}\left[Q_{JAS99} \leq \hat{\theta}_{JAS,0.333}\right] = 0.333$

$\hat{P}\left[Q_{JAS99} > \hat{\theta}_{JAS,0.667}\right] = 0.333$

$\hat{P}\left[T_{ASO99} \leq \hat{\tau}_{ASO,0.333}\right] = 0.333$

$\hat{P}\left[T_{ASO99} > \hat{\tau}_{ASO,0.667}\right] = 0.333$

$\hat{P}\left[Q_{ASO99} \leq \hat{\theta}_{ASO,0.333}\right] = 0.333$

$\hat{P}\left[Q_{ASO99} > \hat{\theta}_{ASO,0.667}\right] = 0.333$

$\hat{P}\left[T_{SON99} \leq \hat{\tau}_{SON,0.333}\right] = 0.393$

$\hat{P}\left[T_{SON99} > \hat{\tau}_{SON,0.667}\right] = 0.273$

$\hat{P}\left[Q_{SON99} \leq \hat{\theta}_{SON,0.333}\right] = 0.333$

$\hat{P}\left[Q_{SON99} > \hat{\theta}_{SON,0.667}\right] = 0.333$

Figure 4-5. Event probability estimate equations for September 1998 Lake Superior basin climate outlook.

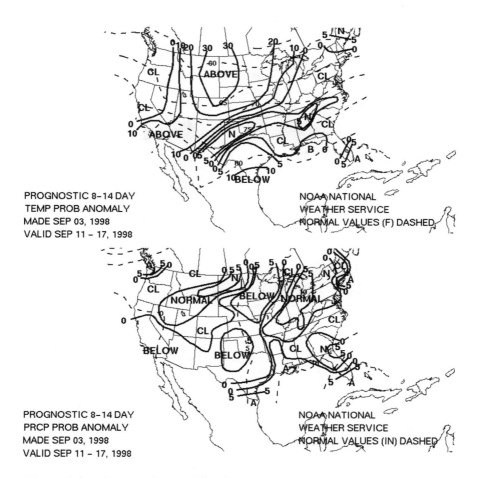

PROGNOSTIC 8–14 DAY
TEMP PROB ANOMALY
MADE SEP 03, 1998
VALID SEP 11 – 17, 1998

NOAA NATIONAL
WEATHER SERVICE
NORMAL VALUES (F) DASHED

PROGNOSTIC 8–14 DAY
PRCP PROB ANOMALY
MADE SEP 03, 1998
VALID SEP 11 – 17, 1998

NOAA NATIONAL
WEATHER SERVICE
NORMAL VALUES (IN) DASHED

Figure 4-6. NOAA 8–14 day probabilistic outlook for September 11-17, 1998.

ties, shown in Figure 4-7. This was downloaded from the CDC Web site. These maps show the forecast probability of being in the upper tercile (above normal) or the lower tercile (below normal). Tercile boundaries are computed for a 31-day period at each grid point, using 1968–97 data. Probabilities are derived from an NCEP ensemble (simulations based on alternative initial conditions, described subsequently), and are calibrated by using past forecasts archived at CDC. As before, the actual CDC product is presented in color with both the lower and the upper tercile probabilities plotted together. They are separated here because gray scales for the two are confusing if plotted together. Reading the example in Figure 4-7 for the state of Florida, the resulting probability estimate equation is

$$\hat{P}\left[Q_{4-10\mathrm{Feb}99} > \hat{\theta}_{4-10\mathrm{Feb},\,0.667}\right] = 0.46 \tag{4-5}$$

NOAA ENSEMBLE EVENT PROBABILITY FORECAST PRODUCTS
It is possible to obtain information on the inherent predictability of a deterministic forecast of the weather by running atmospheric models from a number of likely initial con-

29

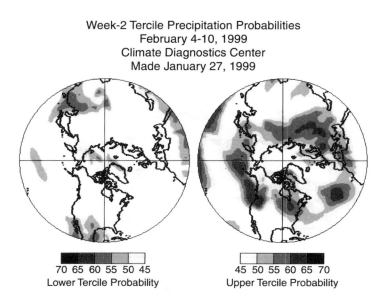

Week-2 Tercile Precipitation Probabilities
February 4-10, 1999
Climate Diagnostics Center
Made January 27, 1999

70 65 60 55 50 45
Lower Tercile Probability

45 50 55 60 65 70
Upper Tercile Probability

Figure 4-7. NOAA week-2 probabilistic outlook for February 4-10, 1999.

ditions, based on actual observations and their likely errors, called an "ensemble" of initial conditions. For a sufficiently large number of realizations (simulations) in the resulting model output ensemble, any forecast quantity can be expressed in terms of probability estimates, conveying information regarding future weather. At NCEP, the ensemble approach has been applied operationally, using the NCEP EMC medium-range forecast model for short-range forecasts as well as using the model in a nonensemble mode for the extended-range *Climate Outlook* previously discussed. These short-range forecasts are created on an experimental basis.

Each day, the EMC publishes on the Web nine successively lagged 1-day outlooks, a 1–5 day outlook, a 6–10 day outlook, and an 8–14 day outlook of precipitation event probabilities. An example is given for the first 24-hour outlook of January 16, 1998, in Figure 4-8. This was downloaded from the NCEP EMC Web site NCEP Ensemble Products of the Global Modeling Branch. The probabilities are given directly in terms of certain specified amounts and do not involve reference to historical quantiles as do NOAA's 1- and 3-month outlooks or their 8–14 day outlooks. Also, no assumptions are being made as to the form of the distributions (NOAA's "four-distribution universe," described previously). The original is in color. Numbers extracted from Figure 4-8 for the Lake Superior basin result in the following probability equations:

$$P\left[P_{12:16\,\text{Jan}\,98-12:17\,\text{Jan}\,98} > 2.54\,\text{mm}\right] \cong 0.35 \tag{4-6a}$$

$$P\left[P_{12:16\,\text{Jan}\,98-12:17\,\text{Jan}\,98} > 6.35\,\text{mm}\right] \cong 0.02 \tag{4-6b}$$

$$P\left[P_{12:16\,\text{Jan}\,98-12:17\,\text{Jan}\,98} > 12.7\,\text{mm}\right] \cong 0.0 \tag{4-6c}$$

$$P\left[P_{12:16\,\text{Jan}\,98-12:17\,\text{Jan}\,98} > 25.4\,\text{mm}\right] \cong 0.0 \tag{4-6d}$$

NOAA Ensemble-based Probability of Precipitation Exceedance
January 16, 12:00 p.m., to January 17, 12:00 p.m., 1998
Made January 16, 1998

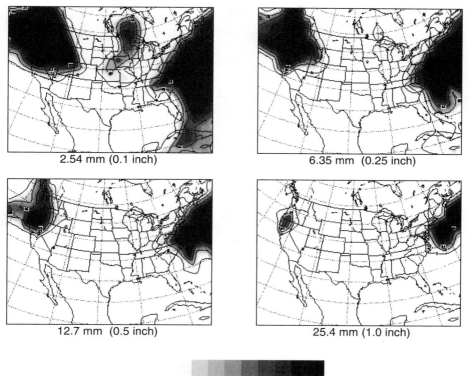

2.54 mm (0.1 inch)

6.35 mm (0.25 inch)

12.7 mm (0.5 inch)

25.4 mm (1.0 inch)

0.0 0.1 0.2 0.3 0.4 0.5 0.6 0.7 0.8 0.9 1.0

Figure 4-8. NOAA 24-hour ensemble forecast product for January 16, 1998.

NOAA 1-DAY PRECIPITATION EVENT PROBABILITY ANOMALY OUTLOOKS
Another NOAA NCEP office, the Hydrometeorology Prediction Center (HPC), makes daily 1-day outlooks for days 3, 4, 5, 6, and 7 into the future for departures from normal of the estimated probability of precipitation. HPC refers to the day-3, 4, and 5 outlooks as medium-range forecast products, which include maps of surface pressure patterns, circulation centers, fronts, daily maximum and minimum temperature anomalies, daily precipitation probability anomalies, and total 5-day precipitation for days 1 through 5. These products are also based on output from the medium-range forecast model and other medium range models and ensembles (simulation runs based on alternative initial conditions). An example for day 3, representing January 30, 1999, made January 27, is given in the top of Figure 4-9. To use it, one requires an estimate of the normal probability of precipitation, which is given in the bottom of Figure 4-9. These products were downloaded from the HPC Web page Current Products and the CPC Web page on Experimental US Threats Assessment. Example numbers are taken from these maps for the Lake Superior basin; the normal probability of precipitation there on January 30 is

United States Day-3 Precipitation Probability Outlook
January 30, 1999, made January 27, 1999
NOAA National Weather Service

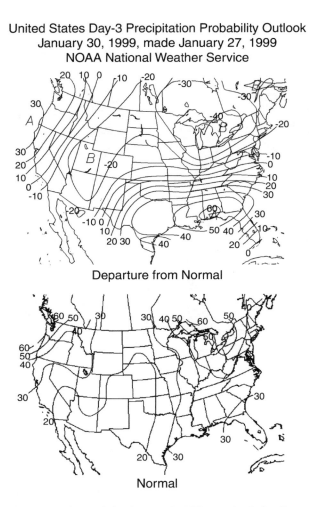

Departure from Normal

Normal

Figure 4-9. NOAA day-3 precipitation probability outlook for January 30, 1999.

about 50%, and the departure from normal is forecast at about −36%. Therefore the outlook for the Lake Superior basin is

$$\hat{P}[Q_{30\text{Jan}99} > 0] = 0.50 - 0.36 = 0.14 \tag{4-7a}$$

$$\hat{P}[Q_{30\text{Jan}99} = 0] = 1 - \hat{P}[Q_{30\text{Jan}99} > 0] = 0.86 \tag{4-7b}$$

EL NIÑO- AND LA NIÑA-BASED EVENT PROBABILITY OUTLOOKS

Consider the following as simply a further example of potentially useful realistic event probability outlooks. The two phases of the El Niño–Southern Oscillation (ENSO) are the El Niño and La Niña events and refer to the oceanic and atmospheric circulation in and over the equatorial Pacific. It is recognized that weather in many parts of the world is related to the occurrence of El Niño and La Niña. In fact, it is one of the prime factors used in NOAA's *Climate Outlook* discussed previously. The study of historical El Niño

Table 4-2. El Niño and La Niña event onset years[a].

Year (1)	Event (2)	Year (3)	Event (4)	Year (5)	Event (6)	Year (7)	Event (8)
1900		1924	La Niña	1948		1972	El Niño
1901		1925	El Niño	1949		1973	La Niña
1902	El Niño	1926	El Niño	1950	La Niña	1974	
1903		1927		1951	El Niño	1975	La Niña
1904	La Niña	1928	La Niña	1952		1976	El Niño
1905	El Niño	1929	El Niño	1953	El Niño	1977	
1906		1930	El Niño	1954		1978	
1907		1931		1955	La Niña	1979	
1908		1932		1956	La Niña	1980	
1909	La Niña	1933		1957	El Niño	1981	
1910	La Niña	1934		1958	El Niño	1982	El Niño
1911	El Niño	1935		1959		1983	
1912	El Niño	1936		1960		1984	
1913		1937		1961		1985	
1914	El Niño	1938	La Niña	1962		1986	El Niño
1915		1939	El Niño	1963		1987	
1916	La Niña	1940		1964	La Niña	1988	La Niña
1917	La Niña	1941	El Niño	1965	El Niño	1989	
1918	El Niño	1942		1966		1990	
1919	El Niño	1943		1967		1991	El Niño
1920		1944		1968		1992	
1921		1945		1969	El Niño	1993	
1922		1946		1970	La Niña	1994	
1923		1947		1971	La Niña		

[a]After Shabbar and Khandekar (1996) and Shabbar et al. (1997).

and La Niña events in Table 4-2 can yield event probabilities useful in hydrology or other derivative outlooks . A simple technique may be applied to derive probabilistic meteorology outlooks that consider the influence of El Niño or La Niña. That is, probabilities of various meteorology events can be estimated from the historical meteorology record *conditioned* on the occurrence of El Niño or La Niña or the absence of both. Then, given that one of these three events is occurring at the time of a forecast, the appropriate set of conditional probabilities can be used as a probabilistic meteorology outlook.

The definition of the occurrence of these events is taken from Shabbar and Khandekar (1996). Strong to moderate ENSO years are defined as those in which the 5-month running Southern Oscillation Index (mean difference in sea-level pressure between Tahiti and Darwin) remained in the lower 25% (El Niño) or upper 25% (La Niña) of the distribution for 5 months or longer. This definition is consistent with that used by Rasmusson (1984), Ropelewski and Jones (1987), and Halpert and Ropelewski (1992). Table 4-2 contains the years of onset of strong or moderate El Niño and La Niña events, as given originally by Shabbar and Khandekar (1996) and corrected and extended by Shabbar et al. (1997).

By inspecting the historical meteorology record for those years of El Niño or La Niña in Table 4-2, one can estimate the probability of any event following an El Niño or La Niña with the event's relative frequency. For example, in the Great Lakes, there is much interest in the effects of ENSO on winter precipitation and air temperatures. Table

Table 4-3. Probability of Lake Superior basin precipitation within historical range[a] given El Niño occurrence.

Range[a] (1)	SON (2)	OND (3)	NDJ (4)	DJF (5)	JFM (6)	FMA (7)	MAM (8)
Lower	0.417	0.583	0.667	0.875	0.667	0.625	0.375
Upper	0.292	0.167	0.042	0.042	0.125	0.167	0.333

[a]Historical ranges are defined as the lowermost third and the uppermost third observed during the 1961–90 period.

Table 4-4. Probability of Lake Superior basin air temperature within historical range[a] given El Niño occurrence.

Range[a] (1)	SON (2)	OND (3)	NDJ (4)	DJF (5)	JFM (6)	FMA (7)	MAM (8)
Lower	0.458	0.375	0.333	0.167	0.208	0.167	0.375
Upper	0.208	0.375	0.500	0.542	0.542	0.417	0.375

[a]Historical ranges are defined as the lowermost third and the uppermost third observed during the 1961–90 period.

4-3 presents the relative frequencies of precipitation over the Lake Superior basin in selected 3-month periods falling within lower and upper thirds of the historical range (observed in 1961–90) following an El Niño. Table 4-4 does the same for air temperatures. *Given* that an El Niño is occurring, the numbers in Tables 4-3 and 4-4 can be interpreted as forecast probabilities *conditioned* on El Niño occurrence and used in making a derivative forecast. For example, in September 1997 it was recognized that a very strong El Niño was occurring. A forecast made at that time for the Lake Superior basin over the December-January-February period would have been, from Tables 4-3 and 4-4,

$$\hat{P}\left[Q_{DJF97} \leq \hat{\theta}_{DJF,\,0.333} \right] \;=\; 0.875 \tag{4-8a}$$

$$\hat{P}\left[Q_{DJF97} > \hat{\theta}_{DJF,\,0.667} \right] \;=\; 0.042 \tag{4-8b}$$

$$\hat{P}\left[T_{DJF97} \leq \hat{\tau}_{DJF,\,0.333} \right] \;=\; 0.167 \tag{4-8c}$$

$$\hat{P}\left[T_{DJF97} > \hat{\tau}_{DJF,\,0.667} \right] \;=\; 0.542 \tag{4-8d}$$

A warm winter is typical in the Great Lakes for an El Niño year, and a dry winter is typical in the upper Great Lakes (Superior, Michigan, Huron, and Georgian Bay) for an El Niño year.

Likewise, Tables 4-5 and 4-6 present Lake Superior basin relative frequencies for La Niña years. Table 4-5 reveals patterns in La Niña winter precipitation (DJF, JFM, and FMA) less consistent than air temperature in Table 4-6. It is also less consistent than either El Niño precipitation or air temperature in Table 4-3 or 4-4, respectively. Therefore, there is little confidence in its use for forecasts. However, winter temperature trends in Table 4-6 are more significant. *Given* that a La Niña year is occurring, the numbers in Tables 4-5 and 4-6 can be interpreted as forecast probabilities *conditioned* on La Niña occurrence in the same manner just described for El Niño. For example, in September 1998, it was recognized that a La Niña was occurring; a forecast made at that time for the Lake Superior basin over the 3-month periods covering October-November-December through January-February-March would have been, from Table 4-6,

$$\hat{P}\left[T_{OND98} \leq \hat{\tau}_{OND,\,0.333} \right] \;=\; 0.412 \tag{4-9a}$$

Table 4-5. Probability of Lake Superior basin precipitation within historical range[a] given La Niña occurrence.

Range[a] (1)	SON (2)	OND (3)	NDJ (4)	DJF (5)	JFM (6)	FMA (7)	MAM (8)
Lower	0.353	0.471	0.353	0.412	0.412	0.412	0.412
Upper	0.235	0.353	0.294	0.353	0.412	0.235	0.176

[a]Historical ranges are defined as the lowermost third and the uppermost third observed during the 1961–90 period.

Table 4-6. Probability of Lake Superior basin air temperature within historical range[a] given La Niña occurrence.

Range[a] (1)	SON (2)	OND (3)	NDJ (4)	DJF (5)	JFM (6)	FMA (7)	MAM (8)
Lower	0.118	0.412	0.588	0.529	0.529	0.412	0.529
Upper	0.294	0.235	0.176	0.235	0.176	0.176	0.118

[a]Historical ranges are defined as the lowermost third and the uppermost third observed during the 1961–90 period.

$$\hat{P}\left[T_{\text{OND98}} > \hat{\tau}_{\text{OND, 0.667}}\right] = 0.235 \tag{4-9b}$$

$$\hat{P}\left[T_{\text{NDJ98}} \leq \hat{\tau}_{\text{NDJ, 0.333}}\right] = 0.588 \tag{4-9c}$$

$$\hat{P}\left[T_{\text{NDJ98}} > \hat{\tau}_{\text{NDJ, 0.667}}\right] = 0.176 \tag{4-9d}$$

$$\hat{P}\left[T_{\text{DJF98}} \leq \hat{\tau}_{\text{DJF, 0.333}}\right] = 0.529 \tag{4-9e}$$

$$\hat{P}\left[T_{\text{DJF98}} > \hat{\tau}_{\text{DJF, 0.667}}\right] = 0.235 \tag{4-9f}$$

$$\hat{P}\left[T_{\text{JFM99}} \leq \hat{\tau}_{\text{JFM, 0.333}}\right] = 0.529 \tag{4-9g}$$

$$\hat{P}\left[T_{\text{JFM99}} > \hat{\tau}_{\text{JFM, 0.667}}\right] = 0.176 \tag{4-9h}$$

A cool winter is typical in the upper Great Lakes area for a La Niña year. See Exercises A2-1, A2-2, and A2-2g in Appendix 2.

NOAA 6–10 DAY MOST-PROBABLE EVENT OUTLOOKS

There are several most-probable event outlook types of interest here. Recall that a most-probable event outlook can be interpreted to indicate that a forecast event has an associated higher than normal probability. Usually, the specification of a single interval (or event) as the most probable presumes that the actual probability density function will be unimodal; i.e., it will have only one peak. Also, most-probable event outlooks are usually interpreted to indicate that all intervals, defined by the issuing agency, other than the indicated most probable, have a smaller than normal probability associated with them.

NOAA's Climate Prediction Center also produces a 6–10 day outlook, covering the 5-day period beginning 6 days in the future. It is issued every few days for both temperature and precipitation events. It predicts which of five intervals of 5-day average air temperature are expected: less than the 10% quantile (much below normal), between the 10% and 30% quantiles (below normal), between the 30% and 70% quantiles (normal), between the 70% and 90% quantiles (above normal), or greater than the 90% quantile (much above normal). The quantiles are defined from observations from 1961 to 1990 (J. D. Hoopingarner, personal communication, 1996). It also predicts which of three in-

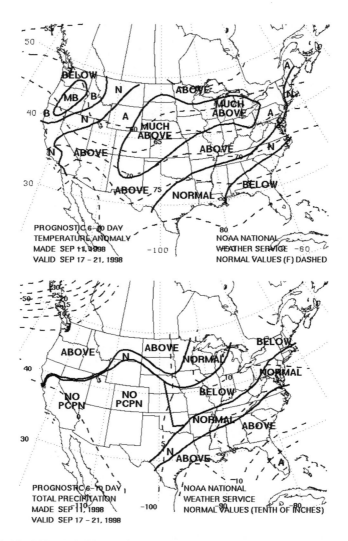

Figure 4-10. NOAA 6–10 day probabilistic outlook for September 17-21, 1998.

tervals (below normal, normal, or above normal) of total precipitation are expected (respectively, the lower, middle, or upper thirds of observations from 1961 to1990) or specifies that no precipitation is expected. Figure 4-10 illustrates NOAA's 6–10 day outlook for September 17–21, 1998, made September 11, 1998. This outlook results in up to eight inequalities. For example, the outlook on the Lake Superior basin is for above-normal air temperatures and normal precipitation, and can be interpreted in terms of probability statements as

$$\hat{P}\left[T_{17-21\text{Sep}98} \leq \hat{\tau}_{17-21\text{Sep}, 0.100} \right] \leq 0.100 \qquad (4\text{-}10\text{a})$$

$$\hat{P}\left[\hat{\tau}_{17-21\text{Sep}, 0.100} < T_{17-21\text{Sep}98} \leq \hat{\tau}_{17-21\text{Sep}, 0.300} \right] \leq 0.200 \qquad (4\text{-}10\text{b})$$

$$\hat{P}\left[\hat{\tau}_{17-21\,\text{Sep},\,0.300} < T_{17-21\,\text{Sep}\,98} \leq \hat{\tau}_{17-21\,\text{Sep},\,0.700}\right] \leq 0.400 \qquad \text{(4-10c)}$$

$$\hat{P}\left[\hat{\tau}_{17-21\,\text{Sep},\,0.700} < T_{17-21\,\text{Sep}\,98} \leq \hat{\tau}_{17-21\,\text{Sep},\,0.900}\right] > 0.200 \qquad \text{(4-10d)}$$

$$\hat{P}\left[T_{17-21\,\text{Sep}\,98} > \hat{\tau}_{17-21\,\text{Sep},\,0.900}\right] \leq 0.100 \qquad \text{(4-10e)}$$

$$\hat{P}\left[Q_{17-21\,\text{Sep}\,98} \leq \hat{\theta}_{17-21\,\text{Sep},\,0.333}\right] \leq 0.333 \qquad \text{(4-10f)}$$

$$\hat{P}\left[\hat{\theta}_{17-21\,\text{Sep},\,0.333} < Q_{17-21\,\text{Sep}\,98} \leq \hat{\theta}_{17-21\,\text{Sep},\,0.667}\right] > 0.334 \qquad \text{(4-10g)}$$

$$\hat{P}\left[Q_{17-21\,\text{Sep}\,98} > \hat{\theta}_{17-21\,\text{Sep},\,0.667}\right] \leq 0.333 \qquad \text{(4-10h)}$$

See Exercises A2-1, A2-2, and A2-2c in Appendix 2. On the other hand, the outlook for the northernmost tip of Texas is for much above normal air temperatures and no precipitation, and can be interpreted in terms of probability statements as

$$\hat{P}\left[T_{17-21\,\text{Sep}\,98} \leq \hat{\tau}_{17-21\,\text{Sep},\,0.100}\right] \leq 0.100 \qquad \text{(4-11a)}$$

$$\hat{P}\left[\hat{\tau}_{17-21\,\text{Sep},\,0.100} < T_{17-21\,\text{Sep}\,98} \leq \hat{\tau}_{17-21\,\text{Sep},\,0.300}\right] \leq 0.200 \qquad \text{(4-11b)}$$

$$\hat{P}\left[\hat{\tau}_{17-21\,\text{Sep},\,0.300} < T_{17-21\,\text{Sep}\,98} \leq \hat{\tau}_{17-21\,\text{Sep},\,0.700}\right] \leq 0.400 \qquad \text{(4-11c)}$$

$$\hat{P}\left[\hat{\tau}_{17-21\,\text{Sep},\,0.700} < T_{17-21\,\text{Sep}\,98} \leq \hat{\tau}_{17-21\,\text{Sep},\,0.900}\right] \leq 0.200 \qquad \text{(4-11d)}$$

$$\hat{P}\left[T_{17-21\,\text{Sep}\,98} > \hat{\tau}_{17-21\,\text{Sep},\,0.900}\right] > 0.100 \qquad \text{(4-11e)}$$

$$\hat{P}\left[Q_{17-21\,\text{Sep}\,98} = 0\right] = 1.000 \qquad \text{(4-11f)}$$

ENVIRONMENT CANADA (EC) MONTHLY MOST-PROBABLE EVENT OUTLOOK

The Environment Canada (EC) Canadian Meteorology Centre (CMC) has been issuing a monthly climate outlook since June 1995, consisting of a most-probable air temperature event. This 1-month outlook is issued twice a month, near the 1st and 15th. It predicts which of three intervals (lower, middle, or upper thirds of observations from 1963 to 1993, respectively below normal, near normal, or above normal) of monthly average air temperature are expected over the Canadian part of North America.

> The predictions are obtained directly from numerical model output, using the CMC's operational Global Spectral Model (Ritchie 1991; Ritchie et al. 1995). The CMC uses simple linear regressions between geopotential thickness (height difference between the 1000 hPa and 500 hPa pressure surfaces), produced by the model, and surface temperature anomalies for each season over the period 1963 to 1993. The CMC uses an ensemble approach to generate probability estimates from the regressions applied to their model, started from their atmospheric analyses taken from satellite and field observations. They also provide maps to help the user define the appropriate temperature values (quantile estimates) to use for definition of the lower, middle, and upper intervals.

Figure 4-11 illustrates Environment Canada's monthly outlook for September 1998, made August 31, 1998. This outlook results in three inequalities. For example, the outlook on Lake Superior is for below normal air temperatures, and is interpreted in terms of probability statements here as

$$\hat{P}\left[T_{\text{Sep}\,98} \leq \hat{\tau}_{\text{Sep},\,0.333}\right] > 0.333 \qquad \text{(4-12a)}$$

$$\hat{P}\left[\hat{\tau}_{\text{Sep},\,0.333} < T_{\text{Sep}\,98} \leq \hat{\tau}_{\text{Sep},\,0.667}\right] \leq 0.334 \qquad \text{(4-12b)}$$

September 1998 Temperature Outlook
Environment Canada
Made August 31, 1998

Figure 4-11. Environment Canada monthly probabilistic outlook for September 1998

$$\hat{P}\left[T_{\text{Sep98}} > \hat{\tau}_{\text{Sep, 0.667}} \right] \leq 0.333 \qquad (4\text{-}12c)$$

(Note again that Lake Superior has been blackened in Figure 4-11 to highlight its location.) See Exercises A2-1, A2-2, and A2-2d in Appendix 2.

EC SEASONAL MOST-PROBABLE EVENT OUTLOOK

The CMC also produces a 3-month seasonal outlook each quarter (in December, March, June, and September) of the most-probable average 3-month air temperature and total precipitation. As with the CMC's monthly outlook, each seasonal outlook predicts which of three intervals of 3-month average air temperature or total precipitation are expected over Canada (lower, middle, or upper thirds of observations from 1963 to 1993 for surface temperature and from 1961 to 1990 for precipitation, respectively below normal, near normal, or above normal).

The technique used to produce the seasonal forecast is similar to the one used to produce the monthly forecast. The main difference is that the seasonal forecast is derived from an ensemble of 12 model runs, with 6 from the CMC global spectral model and 6 from a general circulation model (GCM) of the atmosphere. Both use the same CMC atmospheric analyses but differ in the way they use the analyzed surface fields. For the global spectral model, no interactive ocean is used, and so the sea surface temperatures and snow cover anomalies observed just prior to the beginning of the forecast are fixed throughout the forecast period and added to the evolving climatology. The ice cover anomalies are relaxed to climatology at the end of the first month of the forecast period. For the GCM, snow cover is a prognostic variable and no special treatment is required, but sea surface temperature and ice cover anomalies are handled as before.

Figure 4-12 illustrates Environment Canada's seasonal fall outlook for September-October-November 1998, made August 31, 1998. This outlook can result in six inequalities. For example, the outlook on the Lake Superior basin is for normal air tem-

September-October-November 1998 Seasonal Outlook
Environment Canada
Made August 31, 1998

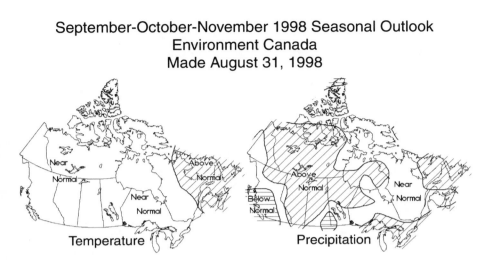

Temperature   Precipitation

Figure 4-12. Environment Canada seasonal fall 1998 probabilistic outlook.

peratures and above normal precipitation, and can be interpreted in terms of probability statements as

$$\hat{P}\left[T_{\text{SON98}} \leq \hat{t}_{\text{SON}, 0.333} \right] \leq 0.333 \qquad (4\text{-}13a)$$

$$\hat{P}\left[\hat{t}_{\text{SON}, 0.333} < T_{\text{SON98}} \leq \hat{t}_{\text{SON}, 0.667} \right] > 0.334 \qquad (4\text{-}13b)$$

$$\hat{P}\left[T_{\text{SON98}} > \hat{t}_{\text{SON}, 0.667} \right] \leq 0.333 \qquad (4\text{-}13c)$$

$$\hat{P}\left[Q_{\text{SON98}} \leq \hat{\theta}_{\text{SON}, 0.333} \right] \leq 0.333 \qquad (4\text{-}13d)$$

$$\hat{P}\left[\hat{\theta}_{\text{SON}, 0.333} < Q_{\text{SON98}} \leq \hat{\theta}_{\text{SON}, 0.667} \right] \leq 0.334 \qquad (4\text{-}13e)$$

$$\hat{P}\left[Q_{\text{SON98}} > \hat{\theta}_{\text{SON}, 0.667} \right] > 0.333 \qquad (4\text{-}13f)$$

See Exercises A2-1, A2-2, and A2-2e in Appendix 2.

EC EXTENDED SEASONAL MOST-PROBABLE EVENT OUTLOOKS
The CMC also produces extended 3-month seasonal outlooks each quarter (in December, March, June, and September) of the most-probable average 3-month air temperature and total precipitation lagged 3 months, 6 months, and 9 months. As with the CMC's regular seasonal outlook, each predicts which of three intervals of 3-month average air temperature or total precipitation are expected over Canada in successive seasons (lower, middle, or upper thirds of observations from 1963 to 1993 for surface temperature and from 1961 to 1990 for precipitation, respectively below normal, near normal, or above normal). Figure 4-13 illustrates Environment Canada's extended seasonal outlooks for DJF 98, MAM 99, and JJA 99, all made August 31, 1998. These outlooks can result in 18 inequalities. For example, on the Lake Superior basin, the DJF 98 and MAM 99 outlooks are for below normal air temperature and normal precipitation; the JJA 99 outlook is for normal air temperature and precipitation. These outlooks can be interpreted in terms of probability statements as shown in Figure 4-14. See Exercises

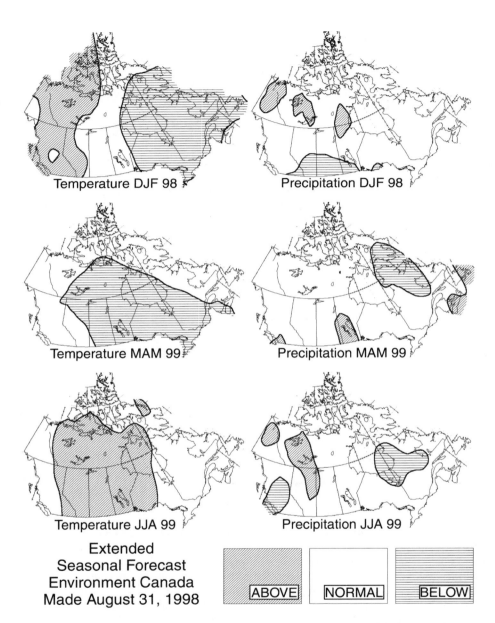

Figure 4-13. Environment Canada extended seasonal probabilistic outlooks.

A2-1, A2-2, and A2-2f in Appendix 2.

OTHER MOST-PROBABLE EVENT OUTLOOKS

All of the outlooks, presented in this chapter, differ in several important respects. They are defined over different time periods (1 day, 5 days, 7 days, 1 month, 3 months). They are defined at different lag times (0, 1, 2, 3, 4, 5, 6, 7, and 8 days and ½, 1½, 2½, 3, 3½,

Lake Superior Basin Most-Probable Event Estimates
DJF 1998 through JJA 1999 Air Temperature & Precipitation
Forecast August 31, 1998 by Environment Canada

$$\hat{P}\left[T_{\text{DJF98}} \leq \hat{\tau}_{\text{DJF}, 0.333} \right] > 0.333$$

$$\hat{P}\left[\hat{\tau}_{\text{DJF}, 0.333} < T_{\text{DJF98}} \leq \hat{\tau}_{\text{DJF}, 0.667} \right] \leq 0.334$$

$$\hat{P}\left[T_{\text{DJF98}} > \hat{\tau}_{\text{DJF}, 0.667} \right] \leq 0.333$$

$$\hat{P}\left[Q_{\text{DJF98}} \leq \hat{\theta}_{\text{DJF}, 0.333} \right] \leq 0.333$$

$$\hat{P}\left[\hat{\theta}_{\text{DJF}, 0.333} < Q_{\text{DJF98}} \leq \hat{\theta}_{\text{DJF}, 0.667} \right] > 0.334$$

$$\hat{P}\left[Q_{\text{DJF98}} > \hat{\theta}_{\text{DJF}, 0.667} \right] \leq 0.333$$

$$\hat{P}\left[T_{\text{MAM99}} \leq \hat{\tau}_{\text{MAM}, 0.333} \right] > 0.333$$

$$\hat{P}\left[\hat{\tau}_{\text{MAM}, 0.333} < T_{\text{MAM99}} \leq \hat{\tau}_{\text{MAM}, 0.667} \right] \leq 0.334$$

$$\hat{P}\left[T_{\text{MAM99}} > \hat{\tau}_{\text{MAM}, 0.667} \right] \leq 0.333$$

$$\hat{P}\left[Q_{\text{MAM99}} \leq \hat{\theta}_{\text{MAM}, 0.333} \right] \leq 0.333$$

$$\hat{P}\left[\hat{\theta}_{\text{MAM}, 0.333} < Q_{\text{MAM99}} \leq \hat{\theta}_{\text{MAM}, 0.667} \right] > 0.334$$

$$\hat{P}\left[Q_{\text{MAM99}} > \hat{\theta}_{\text{MAM}, 0.667} \right] \leq 0.333$$

$$\hat{P}\left[T_{\text{JJA99}} \leq \hat{\tau}_{\text{JJA}, 0.333} \right] \leq 0.333$$

$$\hat{P}\left[\hat{\tau}_{\text{JJA}, 0.333} < T_{\text{JJA99}} \leq \hat{\tau}_{\text{JJA}, 0.667} \right] > 0.334$$

$$\hat{P}\left[T_{\text{JJA99}} > \hat{\tau}_{\text{JJA}, 0.667} \right] \leq 0.333$$

$$\hat{P}\left[Q_{\text{JJA99}} \leq \hat{\theta}_{\text{JJA}, 0.333} \right] \leq 0.333$$

$$\hat{P}\left[\hat{\theta}_{\text{JJA}, 0.333} < Q_{\text{JJA99}} \leq \hat{\theta}_{\text{JJA}, 0.667} \right] > 0.334$$

$$\hat{P}\left[Q_{\text{JJA99}} > \hat{\theta}_{\text{JJA}, 0.667} \right] \leq 0.333$$

Figure 4-14. **Most-probable event estimate equations for extended seasonal Lake Superior basin climate outlook.**

4½, 5½, 6, 6½, 7½, 8½, 9, 9½, 10½, 11½, and 12½ months from when they are issued; real lags depend on when they are actually used). And they specify either a probability of falling within an interval (event probability) or simply the most-probable interval (most-probable event). In the examples presented here (excluding the example El Niño and La Niña outlooks), it is possible on any given day to have as many as 145 equations representing probabilistic meteorology outlooks. More are on the way. Probabilistic meteorology outlooks exist for Africa and Great Britain. Besides those outlooks presented in this book, numerous outlooks provided by other agencies are available now or have been available recently but may be discontinued now. Certainly the available outlooks will constantly be changing with time. However, the examples listed in this section should provide guidance in interpreting other outlooks yet to come.

Chapter 5

DERIVATIVE OUTLOOKS

DERIVATIVE DETERMINISTIC OUTLOOKS

To make meteorology outlooks useful, users must interpret them in terms of their own decisions. That is, they must be able to make derivative forecasts of variables of interest to them. How does one make a hydrology or other derivative outlook from a meteorology outlook? The methods in the category considered here use meteorology outlooks as inputs to hydrology (or other derivative model) transformations. A deterministic method uses the present hydrologic state as initial conditions in a model simulation and then transforms the (deterministic) meteorology outlook to a hydrology outlook with the pertinent model. Figure 5-1 illustrates this method for a watershed model example and a forecast made September 1 for September. The basin moisture at the end of August is 18 cm and becomes the initial condition for a model simulation beginning September 1. The deterministic meteorology forecasts of average September air temperature (5°C) and total September precipitation (6 cm) are used as inputs to the hydrology models for the watershed. Model outputs of basin moisture at the end of September (21 cm) and average September basin runoff (3 cm) form the deterministic derivative hydrology outlook. This method's disadvantage is that the hydrology outlook's skill (ability to accurately forecast) is tied to the low skill of the (deterministic) meteorology outlook.

The derivative model transformation may be one of a wide variety, depending on the problem at hand. For many hydrologists, the transformation may be based on application of the "unit hydrograph," derived from considerations of past storm and flow data for a watershed; see, for example, Linsley et al. (1958). Or it could be based on a solution to the equations of motion for a complex surface representation. The details of the transformation are really unimportant here because the only concern is with the methodology of its use. The transformation models used in later examples will be briefly discussed in those examples. For purposes of illustration at this point, the generic watershed model example of Figure 5-1 and a simple snow accumulation and melt model, following, will be used.

In the simple snow accumulation and melt model example used here, if the daily air temperature is not above 0°C, precipitation will be regarded as snow and there will be no melt (runoff):

$$D_i \quad = \quad D_{i-1} + Q_i \qquad\qquad T_i \leq 0°C \qquad (5-1)$$

Moisture$_{Aug}$ = 18 cm

Temp$_{Sep}$= 5°C | Hydrology Models | Moisture$_{Sep}$ = 21 cm
Prec$_{Sep}$ = 6 cm | | Runoff$_{Sep}$ = 3 cm

Figure 5-1. Example deterministic derivative outlook.

Snowmelt Model

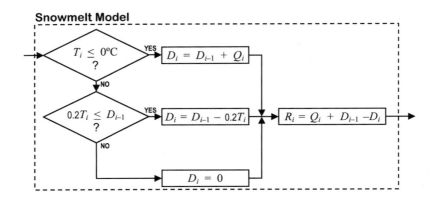

Examples for day 1 with 1.2 cm at the beginning:

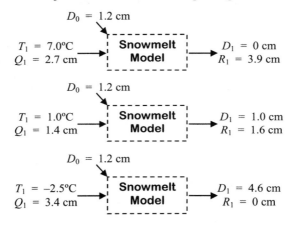

Figure 5-2. Example deterministic derivative (snowmelt) outlooks.

where D_i, Q_i, and T_i are, respectively, end-of-day snow water equivalent, daily precipitation, and daily air temperature, all for day i. If the temperature is above 0°C, any precipitation will be regarded as rainfall and a degree-day snowmelt model will give melt:

$$D_i = D_{i-1} - a T_i \qquad a T_i \leq D_{i-1} \text{ and } T_i > 0°C \qquad (5\text{-}2a)$$

$$D_i = 0 \qquad a T_i > D_{i-1} \text{ and } T_i > 0°C \qquad (5\text{-}2b)$$

where a is the degree-day snowmelt coefficient for a watershed, taken here as 0.2 cm per Celsius degree day. Runoff for day i, R_i, is given from simple continuity on the snow pack, ignoring evaporation and seepage:

$$R_i = Q_i + D_{i-1} - D_i \qquad (5\text{-}3)$$

A second deterministic example is given by using the derivative transformation of Equations (5-1), (5-2), and (5-3) as shown in the top half of Figure 5-2, for the multiple deterministic meteorology forecasts shown in the bottom half of Figure 5-2. There is a single 1-day derived snowmelt forecast for each of three alternative deterministic mete-

44

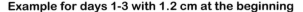

Example for days 1-3 with 1.2 cm at the beginning

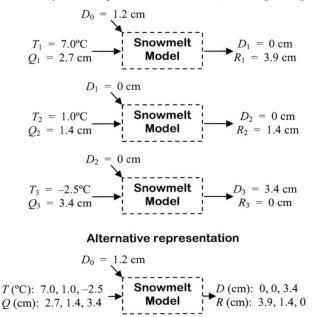

$D_0 = 1.2$ cm

$T_1 = 7.0°C$
$Q_1 = 2.7$ cm
→ Snowmelt Model →
$D_1 = 0$ cm
$R_1 = 3.9$ cm

$D_1 = 0$ cm

$T_2 = 1.0°C$
$Q_2 = 1.4$ cm
→ Snowmelt Model →
$D_2 = 0$ cm
$R_2 = 1.4$ cm

$D_2 = 0$ cm

$T_3 = -2.5°C$
$Q_3 = 3.4$ cm
→ Snowmelt Model →
$D_3 = 3.4$ cm
$R_3 = 0$ cm

Alternative representation

$D_0 = 1.2$ cm

T (°C): 7.0, 1.0, –2.5
Q (cm): 2.7, 1.4, 3.4
→ Snowmelt Model →
D (cm): 0, 0, 3.4
R (cm): 3.9, 1.4, 0

Figure 5-3. Example deterministic derivative (snowmelt) time series outlook.

orology forecasts. A third deterministic example is given in Figure 5-3 by using the inputs in Figure 5-2 arranged chronologically. That is, Figure 5-3 depicts a single 3-day time series forecast for meteorology transformed into a single 3-day time series forecast for snowmelt. The top of Figure 5-3 represents the transformation applied successively to each day's data, generating outputs at the end of the day that are used in the next day's transformation. The alternative representation at the bottom of Figure 5-3 represents the transformation more simply as being applied to the two input time series at the left to generate a single derived snowmelt forecast of the two output time series at the right. A fourth deterministic example is depicted generically in Figure 5-4. It begins by using the present hydrologic state as initial conditions (not pictured) in the derivative model simulations in the same manner as the last examples. Then it uses forecast meteorology time series segments for several variables (two are shown in the figure) as input to the derivative hydrology models for the watershed and transforms them into the corresponding output hydrology forecast time series segments for several other variables (again, two are shown in the figure). Multiple forecasts are made therein.

In Figures 5-2 and 5-4 there are multiple hydrology forecasts; Figure 5-2 depicts multiple single-valued outlooks for two variables, and Figure 5-4 depicts multiple multiple-valued (time series) outlooks for several (two) variables. For these two examples, the "best" outlook must be selected in some manner. The disadvantage here is that there is still low skill in the (alternative) meteorology outlooks, and the problem remains of selecting the best derivative hydrology outlook.

Ordinarily, there is only a single deterministic meteorology forecast (be it a single value or a time series) for use in making a derivative forecast. Yet Figures 5-2 and 5-4

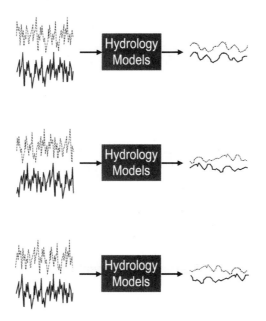

Figure 5-4. Example multiple deterministic derivative outlooks.

depict alternative multiple deterministic meteorology forecasts (of several variables) for input to the derivative transformation models. In practice, multiple deterministic meteorology outlooks are often selected as historical record segments or as simulations from time series (generator) models, which resemble the actual deterministic meteorology forecast in some manner. (For the example meteorology forecasts of Figure 5-1, all historical record segments that begin in September and have September air temperature "close to" 5°C and September precipitation "close to" 6 cm could be used to make multiple time series outlooks as in Figure 5-4.) The consideration of alternative deterministic outlooks is a first step toward making derivative forecasts that recognize multiple possible "futures."

DERIVATIVE PROBABILISTIC OUTLOOKS

Sample Building

How does one make a hydrology outlook from a meteorology outlook that realistically recognizes multiple possibilities? The previous chapter presented many examples of actual meteorology outlooks that embody the concept of recognizing multiple possibilities for the future. The forecasting agencies did this by forecasting probabilities for future possibilities. Users of these meteorology outlooks can interpret the forecast probabilities in terms of the impacts on themselves through "operational hydrology" approaches. Possibilities for the future (multiple deterministic "meteorology scenarios") are identified that resemble past meteorology (preserving observed spatial and temporal relationships) yet are compatible (in some way) with the meteorology outlooks. This was depicted in Figure 5-4 and also is diagrammed in Figure 5-5. In both of these figures, each meteorology scenario actually consists of time series for several meteorology variables. One operational hydrology approach considers historical meteorology as pos-

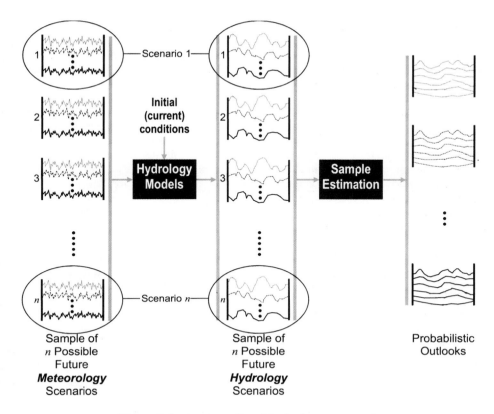

Figure 5-5. An operational hydrology approach.

sibilities for the future (multiple deterministic meteorology scenarios) by segmenting the historical record and using each segment with models and current initial conditions to simulate a hydrology possibility for the future ("hydrology scenario") as in Figures 5-4 and 5-5. Each segment of the historical record then has associated time series of meteorology and hydrology variables, representing a possible scenario for the future, as in Figure 5-5. The approach can then be to consider the resulting set of possible future scenarios as a random sample and infer probabilities and other parameters associated with both meteorology and hydrology through estimation from this sample as defined in Chapter 2 (Croley 1993; Croley and Lee 1993; Croley and Hartmann 1990; Day 1985; Smith et al. 1992). See Exercises A2-1 and A2-3 in Appendix 2. Another operational hydrology approach uses time series models of the meteorology to generate a random sample. This increases the precision of the resulting estimates, because large samples can be generated, but not the accuracy. Use of the historical record to build a random sample directly for estimation avoids the loss of representation consequent with the use of time series models (engendered via selection of an imperfect model and estimation of its parameters from the historical record), but requires a sufficiently large historical record.

The methodology of the operational hydrology approach depicted in Figure 5-5 can be formalized in the following steps. (1) Use the present hydrologic state as initial conditions in derivative model simulations. (2) Select meteorology time series segments as possibilities for the future (meteorology scenarios). These could be, as already mentioned, segments of the historical meteorology record or simulations from appropriate meteorology time series models. Also, each meteorology scenario can actually consist of multiple time series for different meteorology variables. (3) Transform the meteorology time series segments into hydrology scenarios with appropriate derivative hydrology models. Each scenario thus has time series of meteorology and hydrology variables over the relevant time period. The first scenario is produced from the first meteorology time series segments, the second scenario is produced from the second meteorology time series segments, and so forth. (4) Consider the set of possible scenarios (of meteorology and hydrology variables) as a random sample. That is, consider any variable of interest in the scenarios as a random sample by building a collection of its values from all of the scenarios. For example, one could assemble all values for the September watershed runoff, one from each scenario, as a random sample of values of September runoff. Or one could assemble all values for the September air temperature, one from each scenario, as a random sample of values of September air temperature. (5) Infer hydrology and meteorology probabilities by estimating with relative frequencies. For example, count the scenarios with September runoff R_{Sep} less than or equal to 3 cm and divide by the total number of scenarios. Count the scenarios with water surface temperature W_{Sep} greater than 4°C and divide by the total. Count the scenarios with September air temperature T_{Sep} less than or equal to the historical one-third quantile $\hat{\tau}_{Sep,0.333}$ and divide by the total. These relative frequencies form estimates of the corresponding probabilities, as in Equations (2-23):

$$\hat{P}\left[R_{Sep} \leq 3 \text{ cm}\right] = \frac{1}{n} \sum_{i \mid (r_{Sep})_i \leq 3 \text{ cm}} 1 \qquad (5\text{-}4a)$$

$$\hat{P}\left[W_{Sep} > 4°C\right] = \frac{1}{n} \sum_{i \mid (w_{Sep})_i > 4°C} 1 \qquad (5\text{-}4b)$$

$$\hat{P}\left[T_{Sep} \leq \hat{\tau}_{Sep,0.333}\right] = \frac{1}{n} \sum_{i \mid (t_{Sep})_i \leq \hat{\tau}_{Sep,0.333}} 1 \qquad (5\text{-}4c)$$

where $(r_{Sep})_i$, $(w_{Sep})_i$, and $(t_{Sep})_i$ denote the ith value of R_{Sep}, W_{Sep}, and T_{Sep}, respectively, in the random sample. These estimates and others like them, made from the random sample assembled from the scenarios of Figure 5-5, are only as good as the inputs and the models. However, it gives an opportunity to use realistic meteorology as inputs. But it does *not* allow matching of any meteorology outlook.

Sample Restructuring
The operational hydrology approach uses sample estimation tools as if the set of possible future scenarios were a single random sample (i.e., the scenarios are independent of each other and equally likely). This means that the relative frequencies of selected meteorology events in this sample are fixed at values different (generally) from those specified in meteorology outlooks, such as those given in Chapter 4. [For example, while Equation (5-4c) gives an estimated probability of September air temperature in the lower third of its historical range, the first equation in the National Oceanic and Atmospheric Agency (NOAA) 1- and 3-month *Climate Outlook* in Figure 4-5 gives a value for September

48

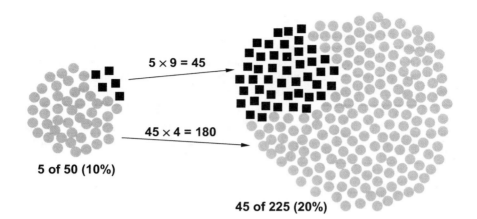

5 × 9 = 45

45 × 4 = 180

5 of 50 (10%)

45 of 225 (20%)

Figure 5-6. Building a biased sample. For example, each square could represent a scenario in which $T_{Sep} > 7°C$ and each circle $T_{Sep} \leq 7°C$

1998 that may be different.] Only by "restructuring" the set of possible future scenarios can one obtain relative frequencies of selected meteorology events that agree with probabilistic meteorology outlooks. This restructuring violates the assumption of independent and equally likely scenarios (no random sample) from the point of view of the meteorology time series segments used. However, the restructured set can be viewed as a random sample of scenarios from the point of view of preserving selected probabilistic meteorology outlooks. There are many methods for restructuring the set of possible future scenarios (Croley 1993; Day 1985; Ingram et al. 1995; Smith et al. 1992).

How may the sample of scenarios be restructured to make a hydrology outlook from a meteorology outlook? A variation on the steps of the operational hydrology approach allows a bias to the approach (the first three steps are the same): (1) use the present hydrologic state as initial conditions in derivative model simulations; (2) select meteorology time series segments as possibilities for the future; (3) transform the meteorology time series segments into hydrology scenarios with appropriate derivative hydrology models (at this point, one can visualize the random sample of scenarios—time series of hydrology and meteorology variables—as a sample of points—each point representing a scenario—in the example on the left half of Figure 5-6); (4) bias the sample (use more of some scenarios than others) to match a meteorology outlook, as in Figure 5-6; and (5) infer derivative probabilities with relative frequencies in the reconstructed sample. Each "item" in Figure 5-6 (each circle or square) represents an entire scenario, consisting of both meteorology and modeled hydrology variable time series segments. The biasing is accomplished by duplicating each scenario in the sample by a duplication count. All of the duplication counts are chosen so that the relative frequency of a desired event matches a desired probabilistic meteorology outlook for that event. For example, suppose only 5 of 50 scenarios (10%) have a September air temperature greater than 7°C (denoted by squares in Figure 5-6) but the probabilistic meteorology outlook says that the probability is 20% that September air temperature will exceed 7°C. To match this, repeat each of the 5 scenarios 9 times and repeat the other 45 scenarios 4 times. This would build a biased sample of 225 scenarios (5 × 9 + 45 × 4) on the right side of Figure 5-6 with 20% (5 × 9 ÷ 225 = 0.2) of the sample with a September air temperature greater than 7°C. Of course, this duplication also applies to all variables in each scenario, hydrology as well as meteorology.

In general, a very large structured set of N possible scenarios (of time series of meteorology and derivative variables) can be constructed by adding (duplicating) each of the available scenarios (in the original set of n possible future scenarios). Each scenario numbered i ($i = 1, 2, \ldots, n$) is duplicated ϑ_i times. By judiciously choosing these duplication numbers ($\vartheta_1, \vartheta_2, \ldots, \vartheta_n$), it is possible to force the relative frequency of any arbitrarily defined group of scenarios in the structured set to any desired value. When this very large structured set is treated as a rándom sample, it gives a relative frequency of (for example) average air temperature or total precipitation (over some time period in the scenarios) satisfying the settings of a desired probabilistic meteorology outlook. For sufficiently large N, desired settings can be approximated to any precision by using integer-valued duplication numbers, ϑ_i. Note also that

$$\sum_{i=1}^{n} \vartheta_i = N \tag{5-5}$$

Building a structured set in this manner to match a desired setting is one of many arbitrary possibilities, but is suggested by considerations of constraints on estimated probability distributions for a single variable, as illustrated shortly in a simple example.

As noted, by treating the N scenarios in the very large structured set as a random sample, one can estimate probabilities and calculate other parameters for all variables. In particular, consider any variable X (either meteorology or hydrology); for example, X might be July-August-September total precipitation, end-of-August soil moisture storage, water surface temperature on day 55, or average June air temperature. Recall from Chapter 2 that the probability of an event is estimated by its relative frequency in the random sample (structured set), which, by Equations (2-23) would be the number of scenarios in which the event occurs divided by the total N:

$$\hat{P}[X \leq x] = \frac{1}{N} \sum_{k \,|\, x_k^N \leq x} 1 \tag{5-6}$$

where x_k^N is the value of variable X for the kth scenario in the very large structured set of N scenarios. For example,

$$\hat{P}[R \leq 3 \text{ cm}] = \frac{1}{N} \sum_{k \,|\, r_k^N \leq 3 \text{ cm}} 1 \tag{5-7a}$$

$$\hat{P}[W > 4°C] = \frac{1}{N} \sum_{k \,|\, w_k^N > 4°C} 1 \tag{5-7b}$$

$$\hat{P}[T \leq \hat{\tau}_{0.333}] = \frac{1}{N} \sum_{k \,|\, t_k^N \leq \hat{\tau}_{0.333}} 1 \tag{5-7c}$$

where R denotes runoff depth, W denotes water surface temperature, and T denotes air temperature, all for some (unspecified) time period. Equations (5-7) are similar to (5-4), but written in terms of the very large structured set constructed from the original sample.

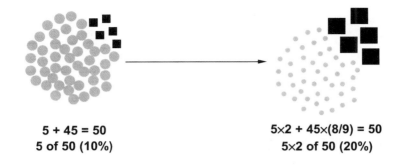

5 + 45 = 50
5 of 50 (10%)

5×2 + 45×(8/9) = 50
5×2 of 50 (20%)

Figure 5-7. Weighting a sample.

Sample Weighting
Note that the sample restructuring can be expressed more generally in terms of *weights* applied to the original sample of scenarios. It is more convenient to work with the original sample of scenarios of size n. For example, consider the duplication of the 5 desired scenarios 9 times and the other 45 scenarios 4 times to build a large sample of 225 (to achieve a 20% relative frequency of desired events) in Figure 5-6. One could instead apply a weight of 2 to the five desired scenarios and 8/9 to the other 45 to get a weighted sample size of 50 with a weighted fraction of 20% of the desired scenarios, as in Figure 5-7. This alternative expression of the biased sample uses the original sample. It makes scenarios of interest a larger percentage of the sample. There is then a weight for each scenario, and probabilities can again be inferred with relative frequencies, expressed in terms of the sample weights w_i :

$$\hat{P}[X \leq x] = \frac{1}{n} \sum_{i \,|\, x_i^n \leq x} w_i \tag{5-8}$$

In more detail, there are actually only n different values of X (x_i^n, $i = 1, \ldots , n$) in the very large structured set, since these n values were duplicated, each by a number ϑ_i, to create the N values. Rewriting Equation (5-6) in terms of the original set of possible future scenarios, for any variable X,

$$\hat{P}[X \leq x] = \sum_{k \,|\, x_k^N \leq x} \frac{1}{N} = \sum_{i \,|\, x_i^n \leq x} \frac{\vartheta_i}{N} = \frac{1}{n} \sum_{i \,|\, x_i^n \leq x} w_i \tag{5-9}$$

or the same result as in Equation (5-8), where

$$w_i = \frac{n}{N} \vartheta_i \tag{5-10}$$

Other Estimators
Likewise, other quantities can be estimated either from the very large structured set or, equivalently, from the original sample by applying weights:

$$\hat{P}\left[X \leq y_j^n\right] = \frac{1}{n} \sum_{l=1}^{j} w_{i(l)} \qquad\qquad j = 1, \ldots , n \tag{5-11a}$$

$$\bar{x} = \frac{1}{n}\sum_{i=1}^{n} w_i x_i^n \tag{5-11b}$$

$$s_X^2 = \frac{1}{n}\sum_{i=1}^{n} w_i \left(x_i^n - \bar{x}\right)^2 \tag{5-11c}$$

$$\widehat{BIAS}_{XZ} = \frac{1}{n}\sum_{i=1}^{n} w_i \left(x_i^n - z_i^n\right) \tag{5-11d}$$

$$\widehat{MSE}_{XZ} = \frac{1}{n}\sum_{i=1}^{n} w_i \left(x_i^n - z_i^n\right)^2 \tag{5-11e}$$

$$\hat{\rho}_{XZ} = \frac{\dfrac{1}{n}\sum_{i=1}^{n} w_i \left(x_i^n - \bar{x}\right)\left(z_i^n - \bar{z}\right)}{\sqrt{s_X^2 s_Z^2}} \tag{5-11f}$$

where y_j^n is defined similarly to y_j, immediately preceding Equation (2-24), as the jth value in a sample of size n ordered from smallest to largest. Likewise, $i(l)$ is the number of the value in the original random sample of size n corresponding to the lth order. Also, x_i^n and z_i^n are values of the random variables X and Z in the original sample of size n. Equation (5-11a) is an estimate of the γ-probability quantile, defined in Equation (2-18), for those values of $\gamma = j/n$, similar to Equation (2-24). Equations (5-11b) through (5-11f) are similar to Equations (2-20). They are estimators of the population characteristics defined in Equations (2-17)—respectively, the mean, variance, bias, mean square error, and correlation.

In more detail, any estimator (defined over the very large structured set of scenarios as if it were a random sample) can be written in terms of the original set. Consider the γ probability quantile for variable X, ξ_γ; recall that Equation (2-18), repeated here, defines it as

$$P\left[X \le \xi_\gamma\right] = \gamma \tag{5-12}$$

It is estimated by the sample order values. In terms of the very large structured set of N scenarios, the γ-probability quantile, ξ_γ, is estimated by the mth order statistic y_m^N for those values of $\gamma = m/N$, where y_m^N is the mth smallest value in a sample of size N, similar to the definitions preceding Equation (2-24). Equation (5-6) gives

$$\hat{P}\left[X \le y_m^N\right] = \frac{m}{N} \qquad m = 1, \ldots, N \tag{5-13}$$

This is similar to Equation (2-24) and can be written similarly to Equation (2-25) as

$$\hat{P}\left[X \le x_{k(m)}^N\right] = \sum_{l=1}^{m} \frac{1}{N} \qquad m = 1, \ldots, N \tag{5-14}$$

where $k(m)$ is the number of the value in the sample of size N corresponding to the mth order. In terms of order statistics for the original set (y_j^n, $j = 1, \ldots, n$), there are $\vartheta_{i(j)}$

identical values of y_j^n in the very large structured set, where $i(j)$ is defined as before [following Equations (5-11)] for the original sample in which $j = 1, \ldots, n$ and $y_j^n = x_{i(j)}^n$. Equations (5-13) and (5-14) may be rewritten in terms of the original set of possible future scenarios (for any variable X):

$$\hat{P}\left[X \le y_j^n\right] = \hat{P}\left[X \le x_{i(j)}^n\right] = \sum_{l=1}^{j} \frac{\vartheta_{i(l)}}{N} \qquad j = 1, \ldots, n \qquad (5\text{-}15)$$

Likewise, the sample mean and variance of variable X over the structured set, $\bar{x}$ and s_X^2, respectively, become, in terms of the original set:

$$\bar{x} = \frac{1}{N} \sum_{k=1}^{N} x_k^N = \frac{1}{N} \sum_{i=1}^{n} \vartheta_i x_i^n \qquad (5\text{-}16a)$$

$$s_X^2 = \frac{1}{N} \sum_{k=1}^{N} \left(x_k^N - \bar{x}\right)^2 = \frac{1}{N} \sum_{i=1}^{n} \vartheta_i \left(x_i^n - \bar{x}\right)^2 \qquad (5\text{-}16b)$$

For a second variable Z (recall that each scenario represents multiple variables), the sample mean $\bar{z}$ and variance s_Z^2 are defined similarly. The sample bias between X and Z, $\widehat{\text{BIAS}}_{XZ}$, the sample mean square error between X and Z, $\widehat{MSE}_{XZ}$, and the sample coefficient of correlation between X and Z, $\hat{\rho}_{XZ}$, become:

$$\widehat{\text{BIAS}}_{XZ} = \frac{1}{N} \sum_{k=1}^{N} \left(x_k^N - z_k^N\right) = \frac{1}{N} \sum_{i=1}^{n} \vartheta_i \left(x_i^n - z_i^n\right) \qquad (5\text{-}17a)$$

$$\widehat{MSE}_{XZ} = \frac{1}{N} \sum_{k=1}^{N} \left(x_k^N - z_k^N\right)^2 = \frac{1}{N} \sum_{i=1}^{n} \vartheta_i \left(x_i^n - z_i^n\right)^2 \qquad (5\text{-}17b)$$

$$\hat{\rho}_{XZ} = \frac{\dfrac{1}{N} \sum_{k=1}^{N} \left(x_k^N - \bar{x}\right)\left(z_k^N - \bar{z}\right)}{\sqrt{s_X^2 s_Z^2}} = \frac{\dfrac{1}{N} \sum_{i=1}^{n} \vartheta_i \left(x_i^n - \bar{x}\right)\left(z_i^n - \bar{z}\right)}{\sqrt{s_X^2 s_Z^2}} \qquad (5\text{-}17c)$$

Equations (5-11) result by substitution of Equation (5-10) into Equations (5-15), (5-16), and (5-17).

Note that by Equations (5-5) and (5-10),

$$\sum_{i=1}^{n} w_i = n \qquad (5\text{-}18)$$

and if all $w_i = 1$, then Equations (5-8) and (5-11) give contemporary (unstructured) estimates from the original set, treated as a random sample. Other statistics can be similarly derived. Now, an advantage is that the biased methodology can be used to match a probabilistic meteorology outlook. Another is that there is more skill in probabilistic meteorology outlooks than in deterministic because the generation of multiple outlooks for estimating probabilities recognizes the possibilities that exist. The disadvantage is that the method is defined for matching only one meteorology event probability in an outlook.

Note that Equation (5-11a) is functionally the same as presented by Smith et al. (1992); in this treatment, the full development of estimator weights, including resampling and empirical distribution material, is presented and extended for other estimators. Smith et al. (1992) used climatic indices from long-range forecasts to set their weights subjectively. In this treatment, the weights are set objectively to match a meteorology outlook probability. The following examples both match a single probability by finding appropriate values of the weights w_i. A more general approach for matching multiple meteorology outlook probabilities is briefly described afterward in this chapter and fully in Chapter 6.

A Simple Example

Consider probability estimates for a single variable that match an existing outlook. For example, suppose the probability for September precipitation is forecast as in Equation (4-1a). It and its converse must sum to unity:

$$\hat{P}\left[Q_{Sep98} \leq \hat{\theta}_{Sep, 0.333} \right] = 0.353 \qquad (5\text{-}19a)$$

$$\hat{P}\left[Q_{Sep98} > \hat{\theta}_{Sep, 0.333} \right] = 0.647 \qquad (5\text{-}19b)$$

The quantile in Equations (5-19) is estimated from historical data from 1961 to 1990 so that:

$$\hat{P}[Q_{Sep} \leq \hat{\theta}_{Sep, 0.333}] = 0.333 \qquad (5\text{-}20)$$

A very large structured set can be constructed, of size N, of scenarios with relative frequencies satisfying Equations (5-19) by duplicating original scenarios taken from the historical record, such that:

$$\frac{N_L}{N} \cong \hat{P}[Q_{Sep} \leq \hat{\theta}_{Sep, 0.333}] = 0.353 \qquad (5\text{-}21a)$$

$$\frac{N - N_L}{N} \cong \hat{P}[Q_{Sep} > \hat{\theta}_{Sep, 0.333}] = 0.647 \qquad (5\text{-}21b)$$

where N_L = number of scenarios with $Q_{Sep} \leq \hat{\theta}_{Sep, 0.333}$. The original sample of n scenarios has n_L scenarios with $Q_{Sep} \leq \hat{\theta}_{Sep, 0.333}$. Each of the n_L scenarios will be duplicated N_L / n_L times. By making the structured set sufficiently large, the approximations in Equations (5-21) can be made as close as desired. In the limit, as the integers N and N_L grow, the approximations in Equations (5-21) approach equalities.

Of the original n scenarios, the ith scenario is repeated ϑ_i times, where

$$\vartheta_i = \frac{N_L}{n_L} \qquad \forall \ i \mid \left(q_{Sep} \right)_i \leq \hat{\theta}_{Sep, 0.333} \qquad (5\text{-}22a)$$

$$\vartheta_i = \frac{N - N_L}{n - n_L} \qquad \forall \ i \mid \left(q_{Sep} \right)_i > \hat{\theta}_{Sep, 0.333} \qquad (5\text{-}22b)$$

where $\left(q_{Sep} \right)_i$ is the September precipitation in scenario i. Note that Equations (5-22) satisfy Equation (5-5). For N sufficiently large, each ratio ϑ_i is always an integer if the probability estimate settings are specified only to a fixed number of digits. Estimators can be written as functions of either the very large structured set $\left(x_1^N, \ldots, x_N^N \right)$, or the "original" set $\left(x_1^n, \ldots, x_n^n \right)$. For example, the structured sample mean and variance, $\bar{x}$ and s_X^2, respectively, are:

$$\bar{x} = \frac{1}{N} \sum_{k=1}^{N} x_k^N = \frac{1}{N} \sum_{i=1}^{n} \vartheta_i x_i^n = \frac{1}{n} \sum_{i=1}^{n} w_i x_i^n \tag{5-23a}$$

$$s_X^2 = \frac{1}{N} \sum_{k=1}^{N} \left(x_k^N - \bar{x} \right)^2 = \frac{1}{N} \sum_{i=1}^{n} \vartheta_i \left(x_i^n - \bar{x} \right)^2 = \frac{1}{n} \sum_{i=1}^{n} w_i \left(x_i^n - \bar{x} \right)^2 \tag{5-23b}$$

where, by Equations (5-21) and (5-22),

$$w_i = \frac{n}{N} \vartheta_i = 0.353 \frac{n}{n_L} \qquad \forall \ i \mid \left(q_{Sep} \right)_i \leq \hat{\theta}_{Sep,\,0.333} \tag{5-24a}$$

$$w_i = \frac{n}{N} \vartheta_i = 0.647 \frac{n}{n - n_L} \qquad \forall \ i \mid \left(q_{Sep} \right)_i > \hat{\theta}_{Sep,\,0.333} \tag{5-24b}$$

Note that Equations (5-24) satisfy Equation (5-18). Note also, for example, that if the period 1961 to 1990 were also the entire historical record, then, by definition, $n_L/n = 0.333$ and $(n - n_L)/n = 0.667$. Therefore,

$$w_i = 0.353/0.333 = 1.06 \qquad \forall \ i \mid \left(q_{Sep} \right)_i \leq \hat{\theta}_{Sep,\,0.333} \tag{5-25a}$$

$$w_i = 0.647/0.667 = 0.97 \qquad \forall \ i \mid \left(q_{Sep} \right)_i > \hat{\theta}_{Sep,\,0.333} \tag{5-25b}$$

Other estimates can be similarly derived. Furthermore, the preceding development can be made for variables besides precipitation and for any period other than "September" without loss of generality. It is also possible to define other very large structured sets based on another probability quantile besides the one used here, 33.3%, and on other systematic manners of duplicating the original scenarios.

Another Simple Example
Consider probability forecasts for a single variable X, that result from matching an existing outlook for another single variable T. By the theorem of total probability in Equation (2-8),

$$P[X \leq x] = P\left[X \leq x | T \leq t \right] P[T \leq t] + P\left[X \leq x | T > t \right] P[T > t] \tag{5-26}$$

If the conditional probabilities $P\left[X \leq x | T \leq t \right]$ and $P\left[X \leq x | T > t \right]$ are estimated from the sample (historical record), then knowing $P[T \leq t]$ (and $P[T > t] = 1 - P[T \leq t]$) from others' forecasts enables the calculation of the forecast probability of $P[X \leq x]$ by Equation (5-26). Equation (5-26) can be written in terms of probability estimates; by using the definition of conditional probability in Equation (2-5) and by replacing with sample counts, $n_{(A)}$ (number of scenarios in the sample for which event A occurs) in the manner of Equation (2-22),

$$\hat{P}[X \le x] = \frac{n_{(X \le x) \cap (T \le t)}}{n_{(T \le t)}} \hat{P}[T \le t] + \frac{n_{(X \le x) \cap (T > t)}}{n_{(T > t)}} \hat{P}[T > t] \qquad (5\text{-}27a)$$

$$\hat{P}[X \le x] = \sum_{i \,|\, x_i^n \le x \text{ and } t_i^n \le t} \frac{1}{n_{(T \le t)}} \hat{P}[T \le t] + \sum_{i \,|\, x_i^n \le x \text{ and } t_i^n > t} \frac{1}{n_{(T > t)}} \hat{P}[T > t] \qquad (5\text{-}27b)$$

$$\hat{P}[X \le x] = \frac{1}{n} \sum_{i \,|\, x_i^n \le x} w_i \qquad (5\text{-}27c)$$

which agrees with Equation (5-8) where

$$w_i = \frac{n}{n_{(T \le t)}} \hat{P}[T \le t] \qquad \forall \; i \,|\, t_i^n \le t \qquad (5\text{-}28a)$$

$$w_i = \frac{n}{n_{(T > t)}} \hat{P}[T > t] \qquad \forall \; i \,|\, t_i^n > t \qquad (5\text{-}28b)$$

Note, while $\hat{P}[X \le x | T \le t]$ and $\hat{P}[X \le x | T > t]$ are unchanged from the historical record estimates, $\hat{P}[X \le x]$ and (of course) $\hat{P}[T \le t]$ are changed from the historical record estimate, reflecting the biasing of the sample to match the forecast of $P[T \le t]$. Furthermore, one can easily show that $\hat{P}[X \le x, T \le t]$ and $\hat{P}[T \le t | X \le x]$ are also changed from the historical record estimates. This reveals an important distinction between the Bayesian statistic approach for estimation of conditional probabilities and the biased sampling approach used here. In the former approach, probabilities of an event A, conditioned on the occurrence of an event B (called *a posterior*, $P[A|B]$) are estimated from an unconditional probability (*a priori*, $P[A]$), a *likelihood* function ($P[B|A]$), both estimated from the record, and an experimental observation (in this case, a forecast, $P[B]$). In terms of estimators, the Bayes theorem is (Pfeiffer 1965):

$$\hat{P}[A|B] = \frac{\hat{P}[A]\hat{P}[B|A]}{\hat{P}[B]} \qquad (5\text{-}29)$$

In the latter approach, one is estimating new joint probabilities (for A and B) that preserve all conditional probabilities observed in the record $P[A|B]$, conditioned on a key event B, while matching that key event (forecast) probability $P[B]$. The theorem of total probability allows one to calculate a new probability for the event of interest (A) as in Equations (2-8) or (5-26). Thus, the biased sample procedure uses conditional probabilities observed in the record rather than estimating them anew from forecasts of selected meteorology events. It estimates the probabilities of events of interest from a new joint distribution that also matches the selected meteorology event forecasts.

STEP-BY-STEP INSTRUCTIONS

The following gives readers a generalized procedure for actually using probabilistic meteorology outlooks in forming their own derivative outlooks. Readers can use the software described in Appendix 2 while following these steps.

1. Select application area(s).
2. Assemble all historical data for meteorology outlook variables over each application area.
3. Select the forecast date and the length of the forecast.
4. Select segments of the historical record to use as possible futures (meteorology time series segments) in the operational hydrology approach, defined by their start date (usually agrees with the forecast date) and beginning year of record.
5. For each application area:
   a. Select the probabilistic meteorology outlooks to use and the relevant probabilities.
   b. Estimate all reference quantiles over the relevant reference periods from the historical data for all of the meteorology outlook probability statements.
   c. Enter probabilities from meteorology outlooks (step a) into their respective probability statements defined in terms of reference quantiles estimated from the historical record (step b).
   d. Check for and eliminate conflicting combinations of probability statements.
   e. Order the probability statements according to importance.
6. Combine all prioritized probability statements for all application areas into a single pool of equations and select from them and reprioritize by reordering (if desired) those to be used subsequently.
7. Identify, for each probability statement, the set of all historical meteorology time series segments containing the statement's defining event and write a corresponding equation on sample weights as in Equation (5-8).
8. To the resulting set of equations, add the constraint of Equation (5-18) as the highest-priority equation, guaranteeing that all weights sum to the number of historical meteorology time series segments.
9. Eliminate redundant equations and any lower-priority equations that make the equation set infeasible (no solution exists).
10. If the resulting nonredundant and feasible equations number the same as the number of historical meteorology time series segments (number of weights), find the simultaneous solution (the weights). Use the weights in Equations (5-11) to make derivative statements for any desired meteorology or hydrology variable (derivative outlooks).
11. If the resulting set of equations does not number the same as the number of weights, it must number fewer than that number (as shown in the next chapter), and multiple solutions for the weights must exist. Pick one of several solutions in some manner to use in making the derivative forecast. The manner of selection is the subject of following chapters.

Readers should now be able to appreciate most of the software described in Appendix 2 and to calculate weights for their own derivative outlooks. The three software modules described in Appendix 2 correspond to these step-by-step instructions as follows: the Simulations Settings module is used for steps 1 through 4; the Climate Outlooks Settings module is used for steps 5a through 5e; and the Mix Outlooks & Compute Weights module is used for steps 6 through 11. Of course, if only a single application area is identified for the problem at hand, then steps 5e and 6 are redundant; the prioritization may then be performed in either the Climate Outlooks Settings module or the Mix Outlooks & Compute Weights module. Perusal of the remaining chapters will enable full appreciation of the many issues involved in using meteorology outlooks, as well as familiarity with the many examples presented there.

PART II

EXTENSIONS AND EXAMPLES

Chapter 6

MULTIPLE METEOROLOGY EVENT PROBABILITIES

Chapter 5 culminated in the example transformation of a probabilistic meteorology out-look, consisting of a single event probability, into a set of sample weights for making a derivative outlook. A single probability from a meteorology event outlook was matched by biasing the operational hydrology sample so that the relative frequency of that event agreed with the forecast probability (e.g., the illustration in Figure 5-6 or 5-7). There are two important ways to extend this transformation. The first is to allow consideration of more than only a single event probability; most of the probabilistic meteorology out-looks actually available, reviewed in Chapter 4, are for multiple event probabilities or can be combined into multiple event probabilities. See, for example, the National Oce-anic and Atmospheric Administration's (NOAA's) 1- and 3-month outlooks (*Climate Outlook*), 8–14 day outlooks, ensemble forecast products, 1-day precipitation probability outlooks, and example El Niño and La Niña precipitation and temperature outlooks in Chapter 4. The second important way to extend the single-event transformation of Chapter 5 is to consider the most-probable event outlook type, in which the meteorology outlook is expressed in terms of one or more most likely events rather than by event probabilities. The latter extension is reserved for Chapter 8. The present chapter con-siders multiple meteorology event probabilities in forming a derivative outlook.

CONSIDERING MULTIPLE OUTLOOKS

How is a probabilistic hydrology outlook made from multiple meteorology outlooks of event probabilities? Equation (5-8) allows the matching of a *single* meteorology event probability,

$$\hat{P}[X \leq x] = \gamma \tag{6-1}$$

where X represents some appropriate meteorology variable in a probabilistic outlook with the associated probability γ. By Equation (5-8), but replacing the sample size no-tation x_i^n with x_i,

$$\frac{1}{n} \sum_{i \mid x_i \leq x} w_i = \gamma \tag{6-2}$$

Recall also Equation (5-18)

$$\sum_{i=1}^{n} w_i = n \tag{6-3}$$

An alternative way of writing Equations (6-2) and (6-3) is

$$\sum_{i=1}^{n} \alpha_i w_i = n\gamma \tag{6-4a}$$

$$\sum_{i=1}^{n} w_i = n \qquad (6\text{-}4b)$$

where

$$\alpha_i = 1 \qquad \forall\; i \mid x_i \le x \qquad (6\text{-}5a)$$

$$\alpha_i = 0 \qquad \forall\; i \mid x_i > x \qquad (6\text{-}5b)$$

An equation of the form of Equation (6-2) can be written similarly to match, for example, each of the 56 meteorology outlook event probabilities in the NOAA 1- and 3-month outlooks [see, for example, Equations (4-2)]:

$$\sum_{i \mid (t_g)_i \le \hat{\tau}_{g,\,0.333}} w_i = a_g n \qquad g = 1,\,\ldots,\,14 \qquad (6\text{-}6a)$$

$$\sum_{i \mid (t_g)_i > \hat{\tau}_{g,\,0.667}} w_i = b_g n \qquad g = 1,\,\ldots,\,14 \qquad (6\text{-}6b)$$

$$\sum_{i \mid (q_g)_i \le \hat{\theta}_{g,\,0.333}} w_i = c_g n \qquad g = 1,\,\ldots,\,14 \qquad (6\text{-}6c)$$

$$\sum_{i \mid (q_g)_i > \hat{\theta}_{g,\,0.667}} w_i = d_g n \qquad g = 1,\,\ldots,\,14 \qquad (6\text{-}6d)$$

where $(t_g)_i$ and $(q_g)_i$ are defined in similar manner to the variables in Equations (5-4) as denoting the ith values of T_g and Q_g, respectively, in the random sample, and where a_g, b_g, c_g, d_g, $\hat{\tau}_{g,\,0.333}$, $\hat{\tau}_{g,\,0.667}$, $\hat{\theta}_{g,\,0.333}$, and $\hat{\theta}_{g,\,0.667}$ are as defined in Equations (4-2). Again, an alternative way of writing Equations (6-6) and (6-3) is

$$\sum_{i=1}^{n} \alpha_{k,i} w_i = e_k \qquad k = 1,\,\ldots,\,57 \qquad (6\text{-}7)$$

where

$$e_k = a_k n \qquad k = 1,\,\ldots,\,14 \qquad (6\text{-}8a)$$
$$e_k = b_{k-14} n \qquad k = 15,\,\ldots,\,28 \qquad (6\text{-}8b)$$
$$e_k = c_{k-28} n \qquad k = 29,\,\ldots,\,42 \qquad (6\text{-}8c)$$
$$e_k = d_{k-42} n \qquad k = 43,\,\ldots,\,56 \qquad (6\text{-}8d)$$
$$e_k = n \qquad k = 57 \qquad (6\text{-}8e)$$

and

$$\alpha_{k,i} = 1 \qquad \forall\; i \mid (t_k)_i \le \hat{\tau}_{k,\,0.333} \qquad k = 1,\,\ldots,\,14 \qquad (6\text{-}9a)$$

$$\alpha_{k,i} = 0 \qquad \forall\; i \mid (t_k)_i > \hat{\tau}_{k,\,0.333} \qquad k = 1,\,\ldots,\,14 \qquad (6\text{-}9b)$$

$$\alpha_{k,i} = 1 \qquad \forall\; i \mid (t_{k-14})_i > \hat{\tau}_{k-14,\,0.667} \qquad k = 15,\,\ldots,\,28 \qquad (6\text{-}9c)$$

$$\alpha_{k,i} = 0 \qquad \forall\; i \mid (t_{k-14})_i \le \hat{\tau}_{k-14,\,0.667} \qquad k = 15,\,\ldots,\,28 \qquad (6\text{-}9d)$$

$$\alpha_{k,i} = 1 \qquad \forall\; i \mid (q_{k-28})_i \le \hat{\theta}_{k-28,\,0.333} \qquad k = 29,\,\ldots,\,42 \qquad (6\text{-}9e)$$

$$\alpha_{k,i} = 0 \qquad \forall\, i \,|\, (q_{k-28})_i > \hat{\theta}_{k-28,\,0.333} \qquad k = 29,\, \ldots,\, 42 \qquad (6\text{-}9f)$$

$$\alpha_{k,i} = 1 \qquad \forall\, i \,|\, (q_{k-42})_i > \hat{\theta}_{k-42,\,0.667} \qquad k = 43,\, \ldots,\, 56 \qquad (6\text{-}9g)$$

$$\alpha_{k,i} = 0 \qquad \forall\, i \,|\, (q_{k-42})_i \le \hat{\theta}_{k-42,\,0.667} \qquad k = 43,\, \ldots,\, 56 \qquad (6\text{-}9h)$$

$$\alpha_{k,i} = 1 \qquad i = 1,\, \ldots,\, n \qquad\qquad\qquad k = 57 \qquad (6\text{-}9i)$$

An equation can also be similarly written for each of the four probabilities in the NOAA 8–14 day outlooks [e.g., Equations (4-4)] or for the single probability of the NOAA CDC week-2 tercile precipitation outlook [e.g., Equation (4-5)]. This is true, too, for the four probabilities in each of the 12 NOAA ensemble forecast products [outlooks for nine 1-day, one 1–5 day, one 6–10 day, and one 8–14 day periods; e.g., Equation (4-6)]. In fact, an equation in the form of Equation (6-2) or, equivalently, of Equation (6-4a), can be written for any outlook event probability or combination of event probabilities that are available. This includes probability estimates derived from sources other than meteorology forecast agencies (e.g., the example El Niño and La Niña outlooks in Chapter 4).

In general, one wants to match an arbitrary number of event probabilities in some or all of the available multiple probabilistic meteorology outlooks. Equations can be written corresponding to all of the desired event probabilities and Equation (6-3) in the form

$$\sum_{i=1}^{n} \alpha_{k,i} w_i = e_k \qquad k = 1,\, \ldots,\, m \qquad (6\text{-}10)$$

where $\alpha_{k,i}$ has the value 1 or 0 corresponding to the inclusion or exclusion, respectively, of each variable in the respective appropriate sets and e_k corresponds to the meteorology event probabilities specified in the appropriate agency outlooks. Note that Equation (6-10) can be written as a vector equation also:

$$\begin{bmatrix} \alpha_{1,1} & \alpha_{1,2} & \cdots & \alpha_{1,n} \\ \alpha_{2,1} & \alpha_{2,2} & \cdots & \alpha_{2,n} \\ \vdots & & & \vdots \\ \alpha_{m,1} & \alpha_{m,2} & \cdots & \alpha_{m,n} \end{bmatrix} \begin{bmatrix} w_1 \\ w_2 \\ \vdots \\ w_n \end{bmatrix} = \begin{bmatrix} e_1 \\ e_2 \\ \vdots \\ e_m \end{bmatrix} \qquad (6\text{-}11)$$

AN EXAMPLE MULTIPLE OUTLOOK EQUATION SET

As an example, suppose the following hypothetical outlooks for September and September-October-November air temperatures and September precipitation are available over a hypothetical forecaster's area of interest (application area):

$$\hat{P}\left[T_{\text{Sep98}} \le \hat{\tau}_{\text{Sep},\,0.40} \right] = 0.30 \qquad (6\text{-}12a)$$

$$\hat{P}\left[T_{\text{Sep98}} > \hat{\tau}_{\text{Sep},\,0.60} \right] = 0.53 \qquad (6\text{-}12b)$$

$$\hat{P}\left[T_{\text{SON98}} \le \hat{\tau}_{\text{SON},\,0.40} \right] = 0.34 \qquad (6\text{-}12c)$$

$$\hat{P}\left[T_{\text{SON98}} > \hat{\tau}_{\text{SON},\,0.60} \right] = 0.56 \qquad (6\text{-}12d)$$

$$\hat{P}\left[Q_{\text{Sep98}} \le \hat{\theta}_{\text{Sep},\,0.40} \right] = 0.36 \qquad (6\text{-}12e)$$

$$\hat{P}\left[Q_{\text{Sep98}} > \hat{\theta}_{\text{Sep},\,0.60} \right] = 0.33 \qquad (6\text{-}12f)$$

Table 6-1. Hypothetical application: Average air temperature and precipitation.

i	Year	$(t_{\text{Sep}})_i$ (°C)	$(t_{\text{SON}})_i$ (°C)	$(q_{\text{SON}})_i$ (mm)
(1)	(2)	(3)	(4)	(5)
1	1991	14.0	7.5	100
2	1992	10.5	7.0	50
3	1993	13.0	5.5	90
4	1994	9.5	6.0	80
5	1995	11.0	4.5	140
6	1996	12.0	5.0	30
7	1997	15.0	8.0	85

For its forecasts in Equation (6-12), the hypothetical climate prediction agency defines its reference quantiles over the period 1991–95. (While this reference period is obviously unrealistic, and in fact this entire example is over-simplified, the example is a good one to illustrate the methodology. Later examples will feature real-world applications after the reader is acquainted with the methodology.) In this example, the forecaster is making a derivative outlook beginning on September 5, 1998. The September and September-October-November meteorology outlooks in Equation (6-12) may be inappropriate, because 5 days have already elapsed. (This issue, of using a meteorology outlook that begins earlier than the derivative outlook, is discussed later in this chapter.)

Assume that the forecaster has assembled daily historical meteorology data over the entire application area and determined the average air temperature over the entire area for the September period for each year of record (1991–97). The forecaster has also determined the areal average temperature and the total precipitation over the entire area for the September-October-November period for each year of record. They are given in Table 6-1. Data from the 1991–95 reference period in Table 6-1 are ordered in Table 6-2, for use in estimating the reference quantiles. From Equation (2-24), the reader can see that the 0.40 and 0.60 quantiles for the September and September-October-November periods are estimated by the second and third smallest values (out of five) for each variable, in Table 6-2, as

$$\hat{t}_{\text{Sep}, 0.40} = 10.5°C \qquad (6\text{-}13a)$$

Table 6-2. Hypothetical Application: Ordered average air temperatures and precipitation 1991–95.

i	i/n	$(t_{\text{Sep}})_i$ (°C)	$(t_{\text{SON}})_i$ (°C)	$(q_{\text{SON}})_i$ (mm)
(1)	(2)	(3)	(4)	(5)
1	0.20	9.5	4.5	50
2	0.40	10.5	5.5	80
3	0.60	11.0	6.0	90
4	0.80	13.0	7.0	100
5	1.00	14.0	7.5	140

$$\hat{\tau}_{Sep, 0.60} = 11.0°C \tag{6-13b}$$

$$\hat{\tau}_{SON, 0.40} = 5.5°C \tag{6-13c}$$

$$\hat{\tau}_{SON, 0.60} = 6.0°C \tag{6-13d}$$

$$\hat{\theta}_{Sep, 0.40} = 80 \text{ mm} \tag{6-13e}$$

$$\hat{\theta}_{Sep, 0.60} = 90 \text{ mm} \tag{6-13f}$$

The forecaster decides to use each year of the historical data beginning on September 5 from 1991 through 1997 as a possible "future" meteorology time series segment in the operational hydrology approach described in Chapter 5. Thus, there are seven values for each variable in the operational hydrology random sample of scenarios and seven weights in Equation (6-10) or (6-11); $n = 7$. The coefficients in Equation (6-10) or (6-11), $\alpha_{k,i}$, have values of 1 or 0 corresponding to the inclusion or exclusion, respectively, of each variable in the sets indicated in the variable subscripts in Equation (6-12). For the first line in Equation (6-12), Table 6-1 shows that the relation $t_{Sep} \leq \hat{\tau}_{Sep, 0.40}$ (or $t_{Sep} \leq 10.5°C$, where t_{Sep} now represents possible "future" values of T_{Sep98}) is satisfied by the following indices: 2 (corresponding to 1992) and 4 (corresponding to 1994). Thus, Equation (6-12a) would be written, as one of the sums in Equation (6-10), as

$$w_2 + w_4 = 0.30 \times 7 \tag{6-14}$$

Expressing Equation (6-14) in vector form, similar to a row equation in Equation (6-11),

$$[0101000] \begin{bmatrix} w_1 \\ w_2 \\ \vdots \\ w_7 \end{bmatrix} = 0.30 \times 7 \tag{6-15}$$

Likewise, Table 6-1 shows that the relation $t_{Sep} > \hat{\tau}_{Sep, 0.60}$ ($t_{Sep} > 11.0°C$) is satisfied by the following indices: 1 (corresponding to 1991), 3, 6, and 7 (corresponding to 1997). Thus, Equation (6-12b) would be written, similar to an equation for a row in Equation (6-11), as

$$[1010011] \begin{bmatrix} w_1 \\ w_2 \\ \vdots \\ w_7 \end{bmatrix} = 0.53 \times 7 \tag{6-16}$$

In fact, as Table 6-1 shows, Equations (6-12), (6-13), and (6-3) become, in the form of Equation (6-11) with Equation (6-3) accounted for first,

$$
\begin{bmatrix}
1111111 \\
0101000 \\
1010011 \\
0010110 \\
1100001 \\
0101010 \\
1000100
\end{bmatrix}
\begin{bmatrix}
w_1 \\
w_2 \\
\vdots \\
w_7
\end{bmatrix}
=
\begin{bmatrix}
1.00 \times 7 \\
0.30 \times 7 \\
0.53 \times 7 \\
0.34 \times 7 \\
0.56 \times 7 \\
0.36 \times 7 \\
0.33 \times 7
\end{bmatrix}
\tag{6-17}
$$

The equations represented by Equation (6-17) must be solved simultaneously to find the weights.

SOLVING THE SET EQUATIONS

Returning now to the general problem, there are m equations [representing $m-1$ meteorology outlook probability equations and Equation (6-3)] in n unknowns (variables, representing weights for each scenario from the historical record) to solve simultaneously, as in Equation (6-10) or (6-11). The equations represented by Equation (6-11) can be manipulated with the usual methods from elementary algebra to simultaneously solve them for the values of the weights that satisfy all of the equations. The Gauss-Jordan method of elimination is a systematic algorithm for solving a system of linear equations. It requires the same number of nonredundant feasible equations as there are variables to be determined, $m = n$. (The case where $m \neq n$ is considered in Chapter 7.) It works by multiplying an equation [a row in the matrix of Equation (6-11) and a corresponding value in the vector on the right-hand side of Equation (6-11)] by an appropriate constant and subtracting the resulting equation from the other equations to eliminate a variable from them. This is successively done until each variable appears in one and only one equation. The resulting system of equations is entirely equivalent to the original system and may be easily solved by inspection, because each variable is isolated.

Consider the following example for illustration:

$$
\begin{aligned}
1x_1 + 2x_2 + 1x_3 &= 9 \tag{6-18a}\\
2x_1 + 3x_2 + 1x_3 &= 14 \tag{6-18b}\\
1x_1 + 1x_2 + 2x_3 &= 7 \tag{6-18c}
\end{aligned}
$$

Multiply Equation (6-18a) by 2 [the ratio of the first coefficients in Equation (6-18b) and Equation (6-18a)] and subtract it from Equation (6-18b), replacing Equation (6-18b). Also multiply Equation (6-18a) by 1 [the ratio of the first coefficients in Equation (6-18c) and Equation (6-18a)] and subtract it from Equation (6-18c), replacing Equation (6-18c). The resulting group is

$$
\begin{aligned}
1x_1 + 2x_2 + 1x_3 &= 9 \tag{6-19a}\\
0x_1 - 1x_2 - 1x_3 &= -4 \tag{6-19b}\\
0x_1 - 1x_2 + 1x_3 &= -2 \tag{6-19c}
\end{aligned}
$$

The first variable has now been eliminated from all equations but the first. Now multiply Equation (6-19b) by -2 [the ratio of the second coefficients in Equation (6-19a) and Equation (6-19b)] and subtract it from Equation (6-19a), replacing Equation (6-19a). Also multiply Equation (6-19b) by 1 [the ratio of the second coefficients in Equation

(6-19c) and Equation (6-19b)] and subtract it from Equation (6-19c), replacing Equation (6-19c):

$$1x_1 + 0x_2 - 1x_3 = 1 \tag{6-20a}$$
$$0x_1 - 1x_2 - 1x_3 = -4 \tag{6-20b}$$
$$0x_1 + 0x_2 + 2x_3 = 2 \tag{6-20c}$$

The second variable has now been eliminated from all equations but the second. Finally, multiply Equation (6-20c) by $-\frac{1}{2}$ [the ratio of the third coefficients in Equation (6-20a) and Equation (6-20c)] and subtract it from Equation (6-20a), replacing Equation (6-20a). Also, multiply Equation (6-20c) again by $-\frac{1}{2}$ [the ratio of the third coefficients in Equation (6-20b) and Equation (6-20c)] and subtract it from Equation (6-20b), replacing Equation (6-20b):

$$1x_1 + 0x_2 + 0x_3 = 2 \tag{6-21a}$$
$$0x_1 - 1x_2 + 0x_3 = -3 \tag{6-21b}$$
$$0x_1 + 0x_2 + 2x_3 = 2 \tag{6-21c}$$

By eliminating the third variable from all equations but the third, all variables are now isolated (only one variable occurs in each equation), so that the solution is immediately seen in Equations (6-21) to be $(x_1, x_2, x_3) = (2, 3, 1)$. Alternatively, Equations (6-18), (6-19), (6-20), and (6-21) can be equivalently written, respectively, in shorthand notation as

$$\left[\begin{array}{ccc|c} 1 & 2 & 1 & 9 \\ 2 & 3 & 1 & 14 \\ 1 & 1 & 2 & 7 \end{array}\right] \rightarrow \left[\begin{array}{ccc|c} 1 & 2 & 1 & 9 \\ 0 & -1 & -1 & -4 \\ 0 & -1 & 1 & -2 \end{array}\right] \rightarrow \left[\begin{array}{ccc|c} 1 & 0 & -1 & 1 \\ 0 & -1 & -1 & -4 \\ 0 & 0 & 2 & 2 \end{array}\right] \rightarrow$$

$$\left[\begin{array}{ccc|c} 1 & 0 & 0 & 2 \\ 0 & -1 & 0 & -3 \\ 0 & 0 & 2 & 2 \end{array}\right] \tag{6-22}$$

Returning to the hypothetical outlooks embodied in Equation (6-17), the solution is given in the steps shown in Figure 6-1, written in the same shorthand as Equation (6-22). From Figure 6-1, the solution is clearly

$$\left(w_1, w_2, w_3, w_4, w_5, w_6, w_7\right) = (1.12, 1.40, 0.77, 0.70, 1.19, 0.42, 1.40) \tag{6-23}$$

USING THE WEIGHTS

With a solution (set of weights) to Equation (6-10) or (6-11), there is then a weight for each of the meteorology time series segments used as a future scenario in the operational hydrology approach to probabilistic hydrology forecasting. The weights can then be used in Equations (5-8) and (5-11) to make probability statements about, or estimates of population characteristics associated with, any variable of interest, including both meteorology and derivative variables.

Continue the example of the hypothetical forecast of Equation (6-12), which resulted in the weights in Equation (6-23). The weights may now be applied to derivative future scenarios corresponding to segments of the historical meteorology record. As an example, use the hypothetical snowmelt model of Equations (5-1) and (5-2), depicted at the top of Figure 5-2, for constructing derivative scenarios beginning on September 5, 1998,

Equation (6-17):

1	1	1	1	1	1	1	7.00
0	1	0	1	0	0	0	2.10
1	0	1	0	0	1	1	3.71
0	0	1	0	1	1	0	2.38
1	1	0	0	0	0	1	3.92
0	1	0	1	0	1	0	2.52
1	0	0	0	1	0	0	2.31

$\rightarrow$

Isolating Variable 1:

1	1	1	1	1	1	1	7.00
0	1	0	1	0	0	0	2.10
0	1	0	1	1	0	0	3.29
0	0	1	0	1	1	0	2.38
0	0	1	1	1	1	0	3.08
0	1	0	1	0	1	0	2.52
0	1	1	1	0	1	1	4.69

$\rightarrow$

Isolating Variable 2:

1	0	1	0	1	1	1	4.90
0	1	0	1	0	0	0	2.10
0	0	0	0	1	0	0	1.19
0	0	1	0	1	1	0	2.38
0	0	1	1	1	1	0	3.08
0	0	0	0	0	1	0	0.42
0	0	1	0	0	1	1	2.59

$\rightarrow$

Isolating Variable 3:

1	0	0	0	0	0	1	2.52
0	1	0	1	0	0	0	2.10
0	0	0	0	1	0	0	1.19
0	0	1	0	1	1	0	2.38
0	0	0	1	0	0	0	0.70
0	0	0	0	0	1	0	0.42
0	0	0	0	-1	0	1	0.21

$\rightarrow$

Isolating Variable 4:

1	0	0	0	0	0	1	2.52
0	1	0	0	0	0	0	1.40
0	0	0	0	1	0	0	1.19
0	0	1	0	1	1	0	2.38
0	0	0	1	0	0	0	0.70
0	0	0	0	0	1	0	0.42
0	0	0	0	-1	0	1	0.21

$\rightarrow$

Isolating Variable 5:

1	0	0	0	0	0	1	2.52
0	1	0	0	0	0	0	1.40
0	0	0	0	0	0	1	1.40
0	0	1	0	0	1	1	2.59
0	0	0	1	0	0	0	0.70
0	0	0	0	0	1	0	0.42
0	0	0	0	-1	0	1	0.21

$\rightarrow$

Isolating Variable 6:

1	0	0	0	0	0	1	2.52
0	1	0	0	0	0	0	1.40
0	0	0	0	0	0	1	1.40
0	0	1	0	0	0	1	2.17
0	0	0	1	0	0	0	0.70
0	0	0	0	0	1	0	0.42
0	0	0	0	-1	0	1	0.21

$\rightarrow$

Isolating Variable 7:

1	0	0	0	0	0	0	1.12
0	1	0	0	0	0	0	1.40
0	0	0	0	0	0	1	1.40
0	0	1	0	0	0	0	0.77
0	0	0	1	0	0	0	0.70
0	0	0	0	0	1	0	0.42
0	0	0	0	1	0	0	1.19

Figure 6-1. Solution Steps for Equation (6-17).

Table 6-3. Hypothetical areal average air temperature (T) and precipitation (Q).

Year	Type	5 Sep.	6 Sep.	7 Sep.	8 Sep.	9 Sep.
(1)	(2)	(3)	(4)	(5)	(6)	(7)
1991	T (°C)	20.0	19.5	17.0	13.0	12.0
	Q (mm)	2.0	8.5	6.0	1.0	3.0
1992	T (°C)	12.0	15.5	15.5	15.0	12.5
	Q (mm)	0.0	1.0	7.0	0.0	10.0
1993	T (°C)	13.5	12.0	11.0	12.0	12.5
	Q (mm)	1.0	0.5	0.0	0.0	1.0
1994	T (°C)	9.0	12.0	14.5	16.5	15.0
	Q (mm)	1.5	3.0	0.0	5.0	9.0
1995	T (°C)	12.0	7.5	10.5	11.5	14.0
	Q (mm)	2.0	4.5	3.5	3.0	5.0
1996	T (°C)	14.0	16.0	17.0	18.5	21.0
	Q (mm)	4.0	9.5	1.0	3.5	0.0
1997	T (°C)	19.5	19.0	17.5	17.5	16.0
	Q (mm)	2.0	0.0	0.0	0.0	1.0

for an application area with initially no snow pack. The first 5 days of each segment of the historical record beginning on September 5 are shown in Table 6-3. Apply the snow melt model to these meteorology time series segments to get the hydrology time series segments shown in Table 6-4. [Note that because all air temperatures in Table 6-3 are above 0°C, Equations (5-1) and (5-2) yield no snow pack buildup, and daily runoff is simply equal to daily precipitation.] Now, the transformed time series in Table 6-4 are weighted by the values in Equation (6-23) and used in Equation (5-11a) to yield the representative probabilistic hydrology outlook estimates in Table 6-5. To illustrate, consider column 3 in Table 6-4 for runoff on September 6, 1998. Table 6-6 details the computations used in applying Equation (5-11a) to make this forecast. In Table 6-6, column 2 is taken from column 3 in Table 6-4 and column 3 is taken from Equation (6-23). Column 2 in Table 6-6 is reordered from smallest to largest in column 4; columns 5 and 6 are the indices and weights corresponding to the entries in column 4, as can be seen by inspection. Column 7 is the running sum of weights in column 6, and column 8 is the running sum in column 7 divided by the number of weights ($n = 7$). Column 8 is thus the nonexceedance probability estimate of Equation (5-11a) for each value in col-

Table 6-4. Hypothetical modeled basin runoff (mm) transformed from Table 6-3 with Equations (5-1) and (5-2).

Year	5 Sep	6 Sep	7 Sep	8 Sep	9 Sep
(1)	(2)	(3)	(4)	(5)	(6)
1991	2.0	8.5	6.0	1.0	3.0
1992	0.0	1.0	7.0	0.0	10.0
1993	1.0	0.5	0.0	0.0	1.0
1994	1.5	3.0	0.0	5.0	9.0
1995	2.0	4.5	3.5	3.0	5.0
1996	4.0	9.5	1.0	3.5	0.0
1997	2.0	0.0	0.0	0.0	1.0

Table 6-5. Hydrology (September 6, 1998, runoff) outlook for Table 6-4 and Equation (6-23) (mm).

Outlook variable (1)	Outlook day, j				
	5 Sep 98 (2)	6 Sep 98 (3)	7 Sep 98 (4)	8 Sep 98 (5)	9 Sep 98 (6)
$\hat{r}_{j,0.15}$	0.00	0.00	0.00	0.00	0.45
$\hat{r}_{j,0.250}$	0.45	0.23	0.00	0.00	0.95
$\hat{r}_{j,0.500}$	1.78	0.98	1.44	0.00	2.63
$\hat{r}_{j,0.750}$	2.00	4.24	5.22	1.94	7.00
$\hat{r}_{j,0.85}$	2.00	6.25	6.25	3.08	9.25

umn 4. (Of course, the example used here is unrealistic, because such a small sample size, $n = 7$, is used and estimates are poor. However, the example serves to illustrate the methodology involved.) Finally, the values in Table 6-5, column 3, are interpolated in column 9 of Table 6-6 from the distribution in column 8. From Table 6-5, for example, the forecast is for 0.75 probability of not exceeding 4.24 mm on September 6, 1998 ($\hat{P}\left[R_{6\text{Sep}98} \leq 4.24 \text{ mm}\right] = 0.75$ or $\hat{r}_{6\text{Sep}98,\,0.75} = 4.24$ mm). (Do not confuse these forecast quantiles with historical quantiles estimated from the historical record.) Stated another way, the forecast probability that runoff on September 6, 1998, will exceed 4.24 mm is 25%.

BAYESIAN FORECASTING

Krzysztofowicz (1999) described a Bayesian Forecasting System (BFS) which incorporates uncertainties from meteorology inputs and hydrology models. It is instructive to place the methods described in this chapter within his framework. In terms of his definitions, the methodology presented here calculates, for a forecast of meteorology probabilities and from a sample of representative events (the historical record), a distribution for the meteorology inputs, represented as vector $\mathbf{W}$, to hydrology models. Krzysztofowicz (1999) referred to this distribution as $\eta(\mathbf{w}|\mathbf{v})$ of $\mathbf{W}$, the generalized probability

---

Table 6-6. Computation of hydrology (September 6, 1998, runoff) outlook quantiles from Table 6-4 and Equation (6-23).

Unordered (original)			Ordered					Interpolations
i	$(r_{6\text{Sep}98})_i$	w_i	$(r_{6\text{Sep}98})_i$	i	w_i	$\sum w_i$	$\frac{1}{n}\sum w_i$	
(1)	(2)	(3)	(4)	(5)	(6)	(7)	(8)	(9)
1	8.5	1.12	0.0	7	1.40	1.40	0.20	$\hat{r}_{6\text{Sep}98,\,0.15} = 0.00$
2	1.0	1.40	0.5	3	0.77	2.17	0.31	$\hat{r}_{6\text{Sep}98,\,0.25} = 0.23$
3	0.5	0.77	1.0	2	1.40	3.57	0.51	$\hat{r}_{6\text{Sep}98,\,0.50} = 0.98$
4	3.0	0.70	3.0	4	0.70	4.27	0.61	
5	4.5	1.19	4.5	5	1.19	5.46	0.78	$\hat{r}_{6\text{Sep}98,\,0.75} = 4.24$
6	9.5	0.42	8.5	1	1.12	6.58	0.94	$\hat{r}_{6\text{Sep}98,\,0.85} = 6.25$
7	0.0	1.40	9.5	6	0.42	7.00	1.00	

density function quantifying the uncertainty about **W** given **v**, the parametric vector specified by a probabilistic quantitative forecast. This notation is inappropriate here as the process described here generates *marginal* distributions satisfying probabilistic meteorology forecasts rather than probabilities conditional on the occurrence of certain parameters **v**. [The biased meteorology sample methodology described here builds a new joint density such that appropriate marginal densities satisfy probabilistic meteorology forecasts; it does not involve calculating conditional distributions, where a distribution is conditioned on the occurrence of an event. For example, the methodology presented here calculates the joint distribution $\phi(\mathbf{w}, \mathbf{v})$ of **W** and **V** such that the (marginal) density of **W** matches meteorology probability forecasts.] Thus the generated marginal distribution of **W** can be denoted simply as $\eta(\mathbf{w})$ of **W**; it can be used in the BFS in the same manner as $\eta(\mathbf{w}|\mathbf{v})$. Krzysztofowicz uses it with the calibrated hydrology model, r, initialized to the present state of the watershed,

$$\mathbf{s} = r(\mathbf{w}, \mathbf{u}) \tag{6-24}$$

where **s** = output vector constituting an estimate of the desired predictand, and **u** = parameter vector comprising all deterministic inputs to the hydrology model, encompassing calibration parameters, model constants or variables known at the time of forecast, and initial conditions. Krzysztofowicz suggests $\eta(\mathbf{w})$ of **W** and $r(\mathbf{w}, \mathbf{u})$ be used to induce density $\pi(\mathbf{s}|\mathbf{u})$ of **S**, quantifying the uncertainty in model output caused by the input uncertainty. This latter density is exactly what results here from the resampled (biased) hydrology scenarios. It then may be further used in a BFS, as described by Krzysztofowicz (1999), to be integrated with hydrology modeling uncertainty to produce an improved hydrology probabilistic outlook.

INCOMPATIBLE OUTLOOK PERIODS

When faced with using a meteorology outlook that begins earlier than the derivative outlook, there are several alternatives. The first alternative is to use such a meteorology outlook but to condition the sample of meteorology time series segments used in the operational hydrology approach on actually observed meteorology data between the start of the meteorology outlook and the start of the derivative outlook. (Do not confuse this with the use of actually observed meteorology data to calculate the initial conditions for hydrology modeling. One must always use available actual meteorology in that manner to update hydrology models to account for current conditions, prior to simulating alternative future scenarios in the operational hydrology approach.) For example, suppose the derivative forecast start date is September 5 and the September (whole month) meteorology outlook is to be used as in the example of Equations (6-12), (6-17), and (6-23) and Tables 6-4 and 6-5]. Because at the time of this forecast actual meteorology has occurred for September 1 through 4, those 4 days of actual meteorology must appear at the beginning of each of the meteorology time series segments used in the operational hydrology approach. This alternative is consistent with the operational hydrology methodology developed here. It recognizes and uses actual meteorology that has occurred within the outlook period for the meteorology forecast in matching the meteorology outlook. However, it may be difficult to match the meteorology outlook with some of the values already fixed in the matching process. Additionally, this alternative supposes that a precondition on meteorology exists (the 4 days of actual observations in this example) that was not used in the generation of the meteorology outlook by the relevant forecast agency. In this example, the forecast agency will have made its September outlook prior to September, without knowledge (of course) of what actually occurred September 1 through 4. Therefore, while the meteorology outlook might be matched in this alternative, there is doubt that the outlook applies, given the occurrence of the actual meteorology. Thus, this alternative is undesirable and is not considered further here.

The second alternative is simply not to use any meteorology outlook that begins earlier than the derivative outlook period. That is, use only meteorology outlooks that begin on or after the start of the derivative outlook. This alternative entails none of the problems enumerated for the first alternative, but it does throw away information that the user might wish to consider. The third alternative is to use a meteorology outlook that begins earlier than the derivative outlook but to construct the sample of meteorology time series segments for use in the operational hydrology approach without using actual meteorology data. (Again, do not confuse this with the use of actual meteorology data to calculate the initial conditions for hydrology modeling.) For example, assume again that the derivative forecast start date is September 5 and that the September (whole month) meteorology outlook is to be used. Even though the actual meteorology data for September 1 through 4 are used to establish the initial conditions for the hydrology models, the first 4 days of meteorology used to build each of the meteorology time series segments comes from the same source as the remainder of the meteorology data. In this example, this source is the historical record. This alternative uses all information in the meteorology outlooks and uses it properly, without the presupposition of observed meteorology, consistent with the meteorology outlooks issued by the forecast agency. Also, it will allow the matching of generally more outlooks than the first alternative. The consequence is that the generated probabilities for meteorology in the derived outlooks (when using *actual* meteorology data) will not match those meteorology outlooks that begin earlier than the derived outlook, since actual meteorology changes these probabilities.

OPERATIONAL HYDROLOGY VARIATIONS
The operational hydrology approach described in Chapter 5 uses available meteorology time series segments (such as historical information) while matching meteorology probability outlooks provided by forecast agencies. Some other approaches severely limit the use of available meteorology time series segments, to be compatible with climate outlooks, or use all segments only by ignoring these outlooks. The National Weather Service is now considering weighting methods that couple historical time series of precipitation with precipitation forecasts (Ingram et al. 1995) for its Extended Streamflow Prediction (ESP) operational hydrology approach (Day 1985; Smith et al. 1992).

Still other approaches use time series models, fitted to historical data, to generate a large sample, increasing precision but not accuracy in the resulting statistical estimates. Direct use of the historical record to build a sample avoids the loss of representation consequent with time series models. In addition, it may not be clear how to modify time series models to agree with climatic outlooks and still be representative of the underlying behavior originally captured in the time series models. Nevertheless, if time series models are used in building the sample, then weighting of this sample, in the manner described in Chapter 5 and in this chapter, to agree with climatic outlooks is straightforward and could still be used.

The determination of the weights involves several choices that were made arbitrarily in this chapter. For example, the weights could be determined directly from multiple climate outlooks, as exemplified in Equations (5-19) through (5-25), or in Equations (5-26) through (5-28), for a single climate outlook. This would involve restrictions on the multiple climate outlooks not considered here.

Efforts can be made to increase the effective sample size beyond the number of years in the historical record. They include the multiple use of record segments or the use of overlapping segments. For example, a 1-month forecast starting on September 5 could use all 1-month historical record segments beginning on September 5 and on September 15. This doubles the sample size, but questions remain on the relevancy of the second-period start date and on the independence of the record segments. It may be that the meteorology 10 days later in the record is still appropriate for that time of year. However, the two periods (September 5 and 15) may not be independent, because they overlap.

Another variation worth noting dispenses with hydrology modeling altogether and simply uses recorded hydrology corresponding to the recorded meteorology in the historical record. This approach ignores the effect of initial conditions in a forecast and requires a sufficiently long hydrological record. It also presupposes unchanged hydrology conditions from the past for the future. However, it may be a suitable approach for a quick forecast approximation in some situations.

SAMPLING AND MODELING INDEPENDENCE

An important advantage associated with the computation of a weighted sample in the operational hydrology approach described in Chapter 5 and in this chapter, and also associated with ESP, is the independence of the weights and the hydrology models. After model simulations are made to build a sample of possible future scenarios for analysis, several probabilistic outlooks can be generated with weights corresponding to the use of different climate outlooks, different methods of considering the climate outlooks, and alternative selections among the outlooks that are available to use each month. In making these alternative analyses and weight (re)computations, it is unnecessary to redo the model simulations to rebuild the sample. This is a real saving when the model simulations are extensive, as is the case with Great Lakes hydrology outlooks. This also enables efficient consideration of other ways of using the weights to make probabilistic outlooks. For example, the use of non-parametric statistics in Equations (5-11) restricts the range of any variable to that present in the historical record or in their hydrology transformations. An alternative that does not restrict range in this manner is to hypothesize a distribution family (e.g., normal, log-normal, log-Pearson type III) and to estimate its moments by utilizing sample statistics defined analogously to those in Equations (5-11). The detractor for parametric estimation is hypothesizing the family of distributions to use.

Most significantly, the method allows joint consideration of multiple meteorology outlooks defined over *different* lengths and periods of time. For the case where there are exactly the same number of unique probabilistic meteorology event probability equations as the number of scenarios to be used from the historical record ($m = n$), the n weights can be calculated directly as shown in the continuing hypothetical example of Equations (6-12), (6-13), (6-17), and (6-23). However, it often happens that the number of unique equations is different from the number of scenarios (and hence the number of weights) in a derivative forecast. This is the subject of the next chapter.

Chapter 7

ALTERNATIVE SOLUTIONS

The previous chapter formulated and solved equations representing user-selected outlooks of meteorology probabilities so that alternative derivative future scenarios could be weighted to reflect those probabilities. The methodology and the continuing example were artificially restricted to the case in which the number of equations [unique meteorology probability outlooks plus Equation (6-3)] m is equal to the number of weights (operational hydrology scenarios) n. Generally, m is not equal to n, and some of the equations may be either redundant or infeasible (nonintersecting with the rest, resulting in no solutions) and must be eliminated.

REDUNDANCY AND INFEASIBILITY

To illustrate in an example, the three equations making up the vector equation

$$\begin{bmatrix} 1111111 \\ 0101000 \\ 1010111 \end{bmatrix} \begin{bmatrix} w_1 \\ w_2 \\ \vdots \\ w_7 \end{bmatrix} = \begin{bmatrix} 1.00 \times 7 \\ 0.30 \times 7 \\ 0.70 \times 7 \end{bmatrix} \tag{7-1}$$

are redundant, because any two of the three can be used to derive the other. For example, adding the second and the third yields the first. Also, the two equations making up the vector equation

$$\begin{bmatrix} 0101000 \\ 0101000 \end{bmatrix} \begin{bmatrix} w_1 \\ w_2 \\ \vdots \\ w_7 \end{bmatrix} = \begin{bmatrix} 0.30 \times 7 \\ 0.40 \times 7 \end{bmatrix} \tag{7-2}$$

are infeasible (their solution does not exist), because there is no point common to both of them and they cannot be solved simultaneously. [Note in this example that subtracting the two equations represented in Equation (7-2) yields an equation easily recognized as impossible, $0 = 0.10 \times 7$]. If m is greater than n, then $m - n$ of the equations *must be* either redundant or infeasible. This corresponds to not being able to simultaneously satisfy all climate outlook information with fewer scenarios than there are outlook boundary conditions.

If there are more equations than variables ($m > n$), then, during the solution process, equations that are redundant or infeasible (nonintersecting with the rest) can be easily identified and eliminated from the system of equations. Consider the earlier example of Equations (6-18) for illustration, where now two additional equations are placed at the end.

$$1x_1 + 2x_2 + 1x_3 = 9 \tag{7-3a}$$

$$2x_1 + 3x_2 + 1x_3 = 14 \qquad (7\text{-}3\text{b})$$
$$1x_1 + 1x_2 + 2x_3 = 7 \qquad (7\text{-}3\text{c})$$
$$2x_1 + 4x_2 + 2x_3 = 18 \qquad (7\text{-}3\text{d})$$
$$3x_1 + 6x_2 + 3x_3 = 30 \qquad (7\text{-}3\text{e})$$

As before, multiply Equation (7-3a) by 2 [the ratio of the first coefficients in Equation (7-3b) and Equation (7-3a)] and subtract it from Equation (7-3b), replacing Equation (7-3b). Also multiply Equation (7-3a) by 1 [the ratio of the first coefficients in Equation (7-3c) and Equation (7-3a)] and subtract it from Equation (7-3c), replacing Equation (7-3c). Now, continuing with Equations (7-3d) and (7-3e), multiply Equation (7-3a) by 2 [the ratio of the first coefficients in Equation (7-3d) and Equation (7-3a)] and subtract the result from Equation (7-3d), replacing Equation (7-3d). Multiply Equation (7-3a) by 3 [the ratio of the first coefficients in Equation (7-3e) and Equation (7-3a)] and subtract it from Equation (7-3e), replacing Equation (7-3e).

$$1x_1 + 2x_2 + 1x_3 = 9 \qquad (7\text{-}4\text{a})$$
$$0x_1 - 1x_2 - 1x_3 = -4 \qquad (7\text{-}4\text{b})$$
$$0x_1 - 1x_2 + 1x_3 = -2 \qquad (7\text{-}4\text{c})$$
$$0x_1 + 0x_2 + 0x_3 = 0 \qquad (7\text{-}4\text{d})$$
$$0x_1 + 0x_2 + 0x_3 = 3 \qquad (7\text{-}4\text{e})$$

The first variable has now been eliminated from all equations but the first. Note that Equation (7-4d) contains no information; it is redundant and can be eliminated. Also note that Equation (7-4e) is an impossible statement; it means that the set of equations is infeasible (has no solution). If it is eliminated also, then the set of (remaining) equations does have a solution as presented in Equation (6-22). Note that both Equations (7-4d) and (7-4e) had all zero coefficients on the left side of the equal sign.

Which equation, in an infeasible set of equations, should be eliminated so that solution of the remaining equations can proceed? When an infeasible set of equations exists, then not all of the equations (and hence not all of the meteorology outlooks they represent) can be simultaneously satisfied. In practice, the forecaster would then have to re-evaluate which meteorology outlooks to use. If the forecaster considers the relative importance of all outlooks at one time, the selection of equations for elimination is better facilitated as the need arises. This can most easily be accomplished by assigning each equation represented by Equation (6-10) or (6-11) a "priority" reflecting its importance to the user. [The highest priority is given to the equation, in the set represented by Equation (6-10) or (6-11), corresponding to Equation (6-3), guaranteeing that all relative frequencies sum to unity.] See Exercises A2-1, A2-2, and A2-2h in Appendix 2 for an easy way to order equations by priority, with the highest priority first. In solving the prioritized system of equations, each equation in order, starting with the next-to-highest priority (second equation in the order), is compared with the set of all higher-priority equations (earlier in the order) and eliminated if it is redundant or is not intersecting (is infeasible). The Gauss-Jordan method of elimination is used both to solve the system of equations and to identify each as redundant or infeasible when compared with all preceding equations; see Figure 7-1. Figure 7-1 embodies the same method of elimination exemplified in Chapter 6, but formalizes it in an algorithm for implementation on the computer. By starting with the higher priorities (earlier orders), each equation is ensured comparison with a known valid set of equations, and higher-priority equations (earlier in

$y_i = i, \quad i = 1, \ldots, n$ ←[variable i is in column i initially]

$z_k = k, \quad k = 1, \ldots, m$ ←[equation k is in row k initially]

$k = 1$ ←[start with row 1]

WHILE $k \leq m$

    $i = k$

    WHILE $i \leq n$ & $\alpha_{z_k, y_i} = 0$ ←[find (next) variable i]

      $i = i + 1$

    IF $i \leq n$ THEN   ←[if feasible and not redundant (i found)]

      IF $i > k$ THEN   ←[rotate columns k through i]

        $x = y_i$

        $y_j = y_{j-1}, \quad j = i, i-1, \ldots, k+1$

        $y_k = x$

      LOOP $i = 1, \ldots, m$ ←[eliminate variable k from other rows]

        IF $k \neq i$ & $\alpha_{z_i, y_k} \neq 0$ THEN

          $r = \alpha_{z_i, y_k} / \alpha_{z_k, y_k}$

          $\alpha_{z_i, j} = \alpha_{z_i, j} - r\,\alpha_{z_k, j}, \quad j = 1, \ldots, n$

          $e_{z_i} = e_{z_i} - r\,e_{z_k}$

    ELSE   ←[eliminate infeasible or redundant row]

      $m = m - 1$

      $z_i = z_{i+1}, \quad i = 1, \ldots, m$

      $k = k - 1$

    $k = k + 1$ ←[prepare to redo new current row]

Gauss-Jordon Elimination Method Applied to Equation (6-11)
(If a row is redundant or infeasible, it is removed from the matrix. Since computations start with row 1, earlier rows are automatically given precedence.)

Figure 7-1. Algorithm for Gauss-Jordan method of elimination.

the order) are kept in preference to lower-priority equations (later in the order). Thus Equation (6-10) or (6-11) can always be reduced so that m is less than or equal to n. If m equals n, then the solution of Equation (6-10) or (6-11) results from Gauss-Jordan elimination as in Figure 7-1 and as exemplified in Chapter 6. If m becomes less than n, then there are multiple solutions to Equation (6-10) or (6-11) and a choice must be made of the one to use.

OPTIMUM SOLUTION

If, after elimination of redundant or infeasible equations, m becomes less than n, then Equation (6-10) or (6-11) consists of the remaining intersecting equations. There are multiple solutions to Equation (6-10) or (6-11) for m less than n, and the identification of the "best" set of weights requires the specification of a measure for comparing the solutions. One such measure is the sum of the deviations of each weight from unity, $\sum (w_i - 1)^2$. (Note: Whenever the preceding sum is written without reference to limits, understand them to be 1 through n.) Solutions of Equation (6-10) or (6-11) that give smaller values of this measure can be judged "better" than those that do not (and the resulting biasing of the original sample of scenarios is minimized in this sense). Other measures are also possible, including those expressing deviation of the weights from a goal, or measures defined on the resulting joint probability distribution function estimates (looking at similarity in joint distributions between the biased and the original sample). For now, it is judged desirable to be as similar to the original sample as is possible, in terms of relative frequencies of the selected events. Alternative measures are considered in Chapter 10.

An optimization problem can be formulated to minimize the deviation of weights from unity in selecting a solution to Equation (6-10) or (6-11):

$$\min \sum_{i=1}^{n} (w_i - 1)^2 \text{ subject to} \tag{7-5a}$$

$$\sum_{i=1}^{n} \alpha_{k,i} w_i = e_k \qquad k = 1, \ldots, m \tag{7-5b}$$

The equation set of Equation (6-10) or (6-11) forms a set of constraints on the minimization of the sum of squared deviations of each weight from unity. The solution of Equations (7-5) may give positive, zero, or negative weights, but only positive or zero (nonnegative) weights make physical sense. Two procedural algorithms enable finding non-negative weights without adding additional constraints to Equations (7-5) (so that the solution is analytically tractable). They repeatedly eliminate the lowest-priority equation in Equations (7-5b) until non-negative weights result. The first procedural algorithm guarantees that strictly *positive* weights result, thereby enabling the use of all possible future scenarios in the operational hydrology sample (no scenario is weighted by zero and effectively eliminated). The second allows *zero* weights, thereby generally allowing more equations in Equations (7-5b) to be used.

Defining the *Lagrangian*, L, for this problem can convert the *constrained* minimization of Equations (7-5) into a single equation whose *unconstrained* minimization is amenable to classical differential calculus (Hillier and Lieberman 1969, pp. 603-08), resulting in a set of linear equations solvable with the Gauss-Jordan method of elimination:

$$L = \sum_{i=1}^{n} (w_i - 1)^2 - \sum_{k=1}^{m} \lambda_k \left(\sum_{i=1}^{n} \alpha_{k,i} w_i - e_k \right) \tag{7-6}$$

where λ_k = the unit penalty of violating the kth constraint in the optimization. Setting the first derivatives of the Lagrangian with respect to each "variable" to zero,

$$\frac{\partial L}{\partial w_i} = 2(w_i - 1) - \sum_{k=1}^{m} \lambda_k \alpha_{k,i} = 0 \qquad i = 1, \ldots, n \tag{7-7a}$$

$$\frac{\partial L}{\partial \lambda_k} = -\sum_{i=1}^{n} \alpha_{k,i} w_i + e_k = 0 \qquad k = 1, \ldots, m \tag{7-7b}$$

gives a set of necessary but not sufficient conditions for the minimization of Equation (7-6) or the problem of Equations (7-5). The solution represents a "critical" point and must be checked further to identify it as either a minimum or a maximum. Equations (7-7) are linear and solvable via the Gauss-Jordan method of elimination because there are $m + n$ equations in $m + n$ unknowns (same number of equations and variables). For this problem [where one of the equations in Equations (6-10) and (7-5) is Equation (6-3)], the solution of Equations (7-7) represents the minimum if $\sum w_i^2 < 2n$ and the maximum if $\sum w_i^2 > 2n$ (see Appendix 3). Note, that Equations (7-7) can be written in vector form as

$$
\begin{bmatrix}
2 & 0 & \cdots & 0 & -\alpha_{1,1} & -\alpha_{2,1} & \cdots & -\alpha_{m,1} \\
0 & 2 & \cdots & 0 & -\alpha_{1,2} & -\alpha_{2,2} & \cdots & -\alpha_{m,2} \\
\vdots & \vdots & & \vdots & \vdots & & & \vdots \\
0 & 0 & \cdots & 2 & -\alpha_{1,n} & -\alpha_{2,n} & \cdots & -\alpha_{m,n} \\
\alpha_{1,1} & \alpha_{1,2} & \cdots & \alpha_{1,n} & 0 & 0 & \cdots & 0 \\
\alpha_{2,1} & \alpha_{2,2} & \cdots & \alpha_{2,n} & 0 & 0 & \cdots & 0 \\
\vdots & \vdots & & \vdots & \vdots & \vdots & & \vdots \\
\alpha_{m,1} & \alpha_{m,2} & \cdots & \alpha_{m,n} & 0 & 0 & \cdots & 0
\end{bmatrix}
\begin{bmatrix}
w_1 \\ w_2 \\ \vdots \\ w_n \\ \lambda_1 \\ \lambda_2 \\ \vdots \\ \lambda_m
\end{bmatrix}
=
\begin{bmatrix}
2 \\ 2 \\ \vdots \\ 2 \\ e_1 \\ e_2 \\ \vdots \\ e_m
\end{bmatrix}
\qquad (7\text{-}8)
$$

For example, take Equation (6-17) without the last line:

$$
\begin{bmatrix}
1111111 \\
0101000 \\
1010011 \\
0010110 \\
1100001 \\
0101010
\end{bmatrix}
\begin{bmatrix}
w_1 \\ w_2 \\ \vdots \\ w_7
\end{bmatrix}
=
\begin{bmatrix}
1.00 \times 7 \\
0.30 \times 7 \\
0.53 \times 7 \\
0.34 \times 7 \\
0.56 \times 7 \\
0.36 \times 7
\end{bmatrix}
\qquad (7\text{-}9)
$$

Now, the number of equations m is less than the number of variables n. Therefore, multiple solutions exist and the optimization of Equations (7-5) can be used. Substituting into Equation (7-8), the following system of linear equations, solved simultaneously through the Gauss-Jordan method of elimination, yields the optimization critical point over the constraint equations represented by Equation (7-9):

$$
\begin{bmatrix}
2 & 0 & 0 & 0 & 0 & 0 & 0 & -1 & 0 & -1 & 0 & -1 & 0 \\
0 & 2 & 0 & 0 & 0 & 0 & 0 & -1 & -1 & 0 & 0 & -1 & -1 \\
0 & 0 & 2 & 0 & 0 & 0 & 0 & -1 & 0 & -1 & -1 & 0 & 0 \\
0 & 0 & 0 & 2 & 0 & 0 & 0 & -1 & -1 & 0 & 0 & 0 & -1 \\
0 & 0 & 0 & 0 & 2 & 0 & 0 & -1 & 0 & 0 & -1 & 0 & 0 \\
0 & 0 & 0 & 0 & 0 & 2 & 0 & -1 & 0 & -1 & -1 & 0 & -1 \\
0 & 0 & 0 & 0 & 0 & 0 & 2 & -1 & 0 & -1 & 0 & -1 & 0 \\
1 & 1 & 1 & 1 & 1 & 1 & 1 & 0 & 0 & 0 & 0 & 0 & 0 \\
0 & 1 & 0 & 1 & 0 & 0 & 0 & 0 & 0 & 0 & 0 & 0 & 0 \\
1 & 0 & 1 & 0 & 0 & 1 & 1 & 0 & 0 & 0 & 0 & 0 & 0 \\
0 & 0 & 1 & 0 & 1 & 1 & 0 & 0 & 0 & 0 & 0 & 0 & 0 \\
1 & 1 & 0 & 0 & 0 & 0 & 1 & 0 & 0 & 0 & 0 & 0 & 0 \\
0 & 1 & 0 & 1 & 0 & 1 & 0 & 0 & 0 & 0 & 0 & 0 & 0
\end{bmatrix}
\begin{bmatrix}
w_1 \\ w_2 \\ w_3 \\ w_4 \\ w_5 \\ w_6 \\ w_7 \\ \lambda_1 \\ \lambda_2 \\ \lambda_3 \\ \lambda_4 \\ \lambda_5 \\ \lambda_6
\end{bmatrix}
=
\begin{bmatrix}
2 \\ 2 \\ 2 \\ 2 \\ 2 \\ 2 \\ 2 \\ 1.00 \times 7 \\ 0.30 \times 7 \\ 0.53 \times 7 \\ 0.34 \times 7 \\ 0.56 \times 7 \\ 0.36 \times 7
\end{bmatrix}
\qquad (7\text{-}10)
$$

The solution of Equation (7-10) is

$$\left(w_1, w_2, w_3, w_4, w_5, w_6, w_7\right) \quad = \quad (1.26, 1.40, 0.77, 0.70, 1.19, 0.42, 1.26) \qquad (7\text{-}11)$$

Because $\sum w_i^2 = 7.8106 \quad (< 2 \times 7)$, the solution minimizes Equation (7-5a).

As mentioned, in general, the solution of Equations (7-5) [or, equivalently, Equation (7-8)] may give positive, zero, or negative weights, but only positive or zero (nonnegative) weights make physical sense, and the optimization must be further constrained to nonnegative weights. This could be done by introducing nonnegativity inequality constraints into Equations (7-5) and into the minimization of the Lagrangian in Equation (7-6). Searching within these additional constraints would require enumeration of all "zero points" or *roots* (a root is a solution with some zero-valued weights). However, this is computationally impractical, because it can involve the inspection of many roots (for example, for $n = 50$, there are potentially $2^{50} - 1$ roots, or more than 10^{15}). (Partial inspection of the many roots is considered in an alternative optimization in Chapter 10.) Furthermore, nonnegativity constraints can result in infeasibility (there is no solution). In this case, additional lowest-priority equations must be eliminated from Equation (6-10) or (6-11) to allow a nonnegative solution. The smallest number possible should be eliminated so that as many as possible of the outlook event probabilities are preserved. Elimination of equations can proceed in a variety of manners. If higher-priority equations were eliminated, it might be possible to eliminate fewer equations. This would involve further assessment of the importance of a small set of high-priority equations versus a larger set of lower-priority equations, which is impossible to make in a general manner for all situations. Therefore, only the successive elimination of lowest-priority equations is used here, one at a time, until a feasible nonnegative solution is found. The following two procedural algorithms provide systematic procedures for finding nonnegative weights by eliminating lowest-priority equations. They also avoid direct use of nonnegativity constraints in Equations (7-5), thus avoiding inspection of the large number of roots that can result (Croley 1996).

The first procedural algorithm solves Equations (7-5) without additional constraints. If the solution also satisfies positivity constraints (all weights are greater than zero), then it is a solution also to the further-constrained optimization problem, and the algorithm is finished. If the solution does not satisfy all of the positivity constraints, then it cannot be an actual solution to the further-constrained problem. This indicates that some positivity constraints would be active in the further-constrained solution, and the constrained optimum may exist only in the limit as some of the weights approach zero (tend to a nonpositive solution). This further-constrained problem is not of interest. Instead, the algorithm removes the lowest-priority equation (reduces m by 1) in the equation set represented by Equation (6-10) or (6-11) and Equations (7-5) and solves the optimization again, repeating until a strictly positive solution results. Figure 7-2 summarizes this procedural algorithm. This procedure guarantees that only *strictly* positive weights result; thus, all possible future scenarios are used (no scenario is weighted by zero and effectively eliminated) in estimating probabilities and other parameters.

Alternatively, if one is willing to disallow some of the possible future scenarios (allow zero weights), then more of the event probabilities in Equation (6-10) or (6-11) can be satisfied in the solution. In the second procedural algorithm, if negative weights are observed in the solution of Equations (7-5), the algorithm adds "zero" constraints ($w_i = 0$) *corresponding only to those weights that are negative* and solves this further-constrained problem. However, introducing selected zero constraints will either eliminate some event probabilities [in Equation (6-10) or (6-11)] (because of infeasibility but

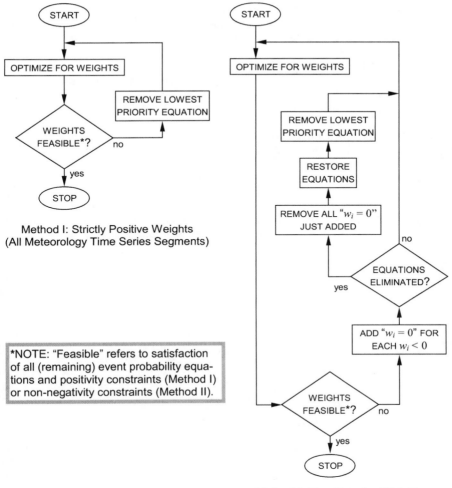

Method I: Strictly Positive Weights
(All Meteorology Time Series Segments)

*NOTE: "Feasible" refers to satisfaction of all (remaining) event probability equations and positivity constraints (Method I) or non-negativity constraints (Method II).

Method II: Non-Negative Weights
(Maximize Use of a Priori Settings)

Figure 7-2. Procedural algorithms for determining physically relevant weights.

not because of redundancy) or it will not. If it does, the solution to the further-constrained problem cannot be feasible in the predecessor problem. The algorithm instead removes the lowest-priority equation in the predecessor problem (reduce m by 1) and re-solves the optimization. If it does not (eliminate some event probabilities), then the optimum solution to the further-constrained problem is feasible (and optimum) in the predecessor problem, but new negative weights could be generated. If no negative weights are generated, then the algorithm is finished. If some negative weights are generated, the algorithm repeats the process (of adding selected zero constraints and solving the further-constrained problem) until either a nonnegative optimum solution is generated to the further-constrained problem or event probabilities are eliminated. If the latter, then the algorithm removes the lowest-priority equation in the predecessor problem (reduces m by 1) and solves the optimization again. This process is repeated until a nonnegative solution results. Figure 7-2 also summarizes this procedural algorithm.

USING THE WEIGHTS

As in Chapter 6, a set of weights are to be used with the meteorology time series segments and their derivative future scenarios to forecast derivative probabilities or to make other probabilistic estimates. Several real-world examples follow.

Hypothetical Snowmelt Revisited

Consider an example similar to the example in Chapter 6, except now use a realistic meteorology forecast and application area data set for the Lake Superior basin. Consider the first pair of equations and the third pair of equations in Figure 4-5 from the National Oceanic and Atmospheric Administration's (NOAA's) 1- and 3-month *Climate Outlook* for September and September-October-November air temperatures over Lake Superior. Add to these the four equations in Equations (4-4) from NOAA's 8–14 day outlook for September 11–17 air temperatures and precipitation over Lake Superior:

$$\hat{P}\left[T_{\text{Sep98}} \le \hat{\tau}_{\text{Sep}, 0.333} \right] = 0.343 \tag{7-12a}$$

$$\hat{P}\left[T_{\text{Sep98}} > \hat{\tau}_{\text{Sep}, 0.667} \right] = 0.323 \tag{7-12b}$$

$$\hat{P}\left[T_{\text{SON98}} \le \hat{\tau}_{\text{SON}, 0.333} \right] = 0.383 \tag{7-12c}$$

$$\hat{P}\left[T_{\text{SON98}} > \hat{\tau}_{\text{SON}, 0.667} \right] = 0.283 \tag{7-12d}$$

$$\hat{P}\left[T_{11-17\text{Sep98}} \le \hat{\tau}_{11-17\text{Sep}, 0.333} \right] = 0.153 \tag{7-12e}$$

$$\hat{P}\left[T_{11-17\text{Sep98}} > \hat{\tau}_{11-17\text{Sep}, 0.667} \right] = 0.513 \tag{7-12f}$$

$$\hat{P}\left[Q_{11-17\text{Sep98}} \le \hat{\theta}_{11-17\text{Sep}, 0.333} \right] = 0.393 \tag{7-12g}$$

$$\hat{P}\left[Q_{11-17\text{Sep98}} > \hat{\theta}_{11-17\text{Sep}, 0.667} \right] = 0.273 \tag{7-12h}$$

In this example, a forecaster will be making a derivative 12-month outlook on September 5, 1998, and feels that the more recent outlooks are most relevant. Therefore, the forecaster subjectively decides to use the more recent September 11–17 air temperature and precipitation forecasts, made September 3, and to throw out the older forecast for September air temperature, made August 13. (Both will be used in a later example.) The forecaster places the September 11–17 forecasts highest in priority and places the extended September-October-November outlook for below-normal air temperature lowest in priority. The forecaster's selection now looks like this:

$$\hat{P}\left[T_{11-17\text{Sep98}} \le \hat{\tau}_{11-17\text{Sep}, 0.333} \right] = 0.153 \tag{7-13a}$$

$$\hat{P}\left[T_{11-17\text{Sep98}} > \hat{\tau}_{11-17\text{Sep}, 0.667} \right] = 0.513 \tag{7-13b}$$

$$\hat{P}\left[Q_{11-17\text{Sep98}} \le \hat{\theta}_{11-17\text{Sep}, 0.333} \right] = 0.393 \tag{7-13c}$$

$$\hat{P}\left[Q_{11-17\text{Sep98}} > \hat{\theta}_{11-17\text{Sep}, 0.667} \right] = 0.273 \tag{7-13d}$$

$$\hat{P}\left[T_{\text{SON98}} \le \hat{\tau}_{\text{SON}, 0.333} \right] = 0.383 \tag{7-13e}$$

$$\hat{P}\left[T_{\text{SON98}} > \hat{\tau}_{\text{SON}, 0.667} \right] = 0.283 \tag{7-13f}$$

where the order of the equations reflects their priorities (highest priority is first). As per the discussion in Chapter 6, Equations (7-13e) and (7-13f) (with periods beginning earlier than September 5) will be imperfectly satisfied with actual rather than historical data.

The forecaster has assembled daily historical meteorology data over the entire Lake Superior basin and determined the average air temperature and the total precipitation over the entire basin for the period September 11–17 for each year of record. The forecaster has also determined the areal average air temperature over the entire basin for the period September-October-November for each year of record. The data are given in Tables 7-1 and 7-2; entries in these tables for other time periods are not used in this example, but are of use in later examples. Recall that both NOAA forecasts (September 11–17, 1998, and September-October-November 1998) are defined relative to historical reference quantiles, estimated over the 1961–90 period (see Chapter 4). Data from the 1961–90 period in Tables 7-1 and 7-2 are ordered in Tables 7-3 and 7-4, respectively, for use in estimating the quantiles. From Equation (2-24), the one-third and two-thirds quantiles for the periods September 11–17 and September-October-November are estimated by the 10th and 20th smallest values (out of 30) for each variable, in Tables 7-3 and 7-4, as

$$\hat{\tau}_{11-17\text{Sep},\,0.333} = 10.38°C \tag{7-14a}$$

$$\hat{\tau}_{11-17\text{Sep},\,0.667} = 12.17°C \tag{7-14b}$$

$$\hat{\theta}_{11-17\text{Sep},\,0.333} = 13.51 \text{ mm} \tag{7-14c}$$

$$\hat{\theta}_{11-17\text{Sep},\,0.667} = 18.83 \text{ mm} \tag{7-14d}$$

$$\hat{\tau}_{\text{SON},\,0.333} = 4.91°C \tag{7-14e}$$

$$\hat{\tau}_{\text{SON},\,0.667} = 5.77°C \tag{7-14f}$$

Again, quantiles for other periods and probabilities in Tables 7-3 and 7-4 are unused in this example, but are of use in later examples. See Exercises A2-1, A2-2, A2-2a, A2-2b, and A2-2h in Appendix 2.

The forecaster decides to use each year of the historical data beginning on September 5 from 1948 through 1994 as a 12-month possible "future" meteorological time series segment in the operational hydrology approach. Thus, there are 47 values for each variable in the operational hydrology random sample of scenarios and 47 weights in Equation (6-10) or (6-11); $n = 47$. The coefficients in Equation (6-10) or (6-11), $\alpha_{k,i}$, have values of 1 or 0 corresponding to the inclusion or exclusion, respectively, of each variable in the sets indicated in the variable subscripts in Equations (7-13). As in the exercise of Equation (6-17) in Chapter 6, inspection of Table 7-1 and 7-2 shows that Equations (7-13), (7-14), and (6-3) become with Equation (6-3) placed first,

$$
\begin{bmatrix}
1111111111111111111111111111111111111111111111 \\
00000110100111001100001001110000100100100000010 \\
10011001011000110011110000001000010000011001101 \\
10000010001110001000110101011100010000000100000 \\
01011101110001000100000010100011101000100011111 \\
00010000000100000110000010101001100001100101110 \\
10000110101010111100001011010001000001100000001
\end{bmatrix}
\begin{bmatrix}
w_1 \\
w_2 \\
\vdots \\
w_{47}
\end{bmatrix}
=
\begin{bmatrix}
1.000 \times 47 \\
0.153 \times 47 \\
0.513 \times 47 \\
0.393 \times 47 \\
0.273 \times 47 \\
0.383 \times 47 \\
0.283 \times 47
\end{bmatrix}
\tag{7-15}
$$

Table 7-1. Average Lake Superior basin air temperatures (°C).

i	Year	$(t_{11-17Sep})_i$	$(t_{17-21Sep})_i$	$(t_{Sep})_i$	$(t_{SON})_i$	$(t_{OND})_i$	$(t_{NDJ})_i$	$(t_{DJF})_i$	$(t_{JFM})_i$
(1)	(2)	(3)	(4)	(5)	(6)	(7)	(8)	(9)	(10)
1	1948	15.37	13.76	14.18	7.25	−0.11	−6.23	−10.35	−10.02
2	1949	11.44	11.06	10.67	5.42	−1.12	−8.94	−12.08	−12.21
3	1950	10.39	10.35	11.47	5.10	−2.70	−9.54	−11.98	−10.11
4	1951	12.91	11.18	10.53	3.57	−3.47	−9.33	−10.55	−9.23
5	1952	16.52	9.92	12.98	5.47	−0.50	−5.38	−8.48	−8.07
6	1953	9.47	11.25	11.89	7.38	0.80	−7.16	−9.52	−9.28
7	1954	9.15	11.34	11.11	5.87	−0.29	−6.24	−10.18	−10.69
8	1955	12.25	14.01	12.06	5.50	−2.35	−8.30	−10.62	−9.44
9	1956	9.91	5.65	9.83	5.91	−0.71	−9.37	−12.57	−11.11
10	1957	13.03	10.23	11.36	5.52	−0.89	−6.30	−10.00	−8.33
11	1958	12.25	12.46	12.47	6.48	−2.30	−10.14	−14.51	−11.66
12	1959	7.99	11.66	13.02	3.74	−2.45	−7.59	−8.78	−10.10
13	1960	9.85	11.81	12.56	6.50	−1.49	−8.46	−11.32	−8.60
14	1961	10.19	15.75	13.05	6.55	−0.85	−8.26	−12.71	−10.87
15	1962	12.68	7.96	11.04	6.27	−0.79	−9.12	−14.12	−12.68
16	1963	12.49	12.22	12.05	8.40	0.30	−6.44	−10.21	−8.58
17	1964	8.77	12.87	10.93	5.12	−2.46	−8.97	−13.16	−12.06
18	1965	8.59	10.30	9.44	4.61	−0.32	−7.74	−10.51	−9.87
19	1966	13.17	12.65	12.43	4.74	−2.64	−8.56	−12.49	−11.15
20	1967	15.50	15.43	12.57	5.04	−1.69	−7.85	−11.17	−9.69
21	1968	15.88	15.95	13.45	6.63	−1.02	−7.54	−10.31	−9.57
22	1969	13.51	11.40	12.29	5.15	−1.44	−8.00	−12.10	−11.94
23	1970	8.67	14.87	12.19	6.16	−1.21	−9.10	−12.49	−11.45
24	1971	12.15	9.42	13.37	7.04	−0.34	−8.84	−13.09	−13.31
25	1972	11.45	11.69	10.24	4.02	−3.63	−8.49	−11.47	−7.33
26	1973	9.79	6.99	11.86	6.47	−0.98	−8.79	−12.92	−11.95
27	1974	10.38	8.18	9.14	4.52	−0.51	−6.08	−9.16	−9.56
28	1975	9.72	11.24	10.37	5.75	−1.19	−8.56	−11.18	−10.01
29	1976	13.48	10.92	11.13	3.51	−5.36	−12.28	−14.22	−9.46
30	1977	11.69	10.81	11.58	5.84	−1.36	−8.45	−12.54	−11.42
31	1978	10.78	11.51	12.31	5.20	−2.95	−10.79	−15.22	−13.01
32	1979	11.92	10.67	11.83	4.67	−1.36	−6.96	−10.40	−10.99
33	1980	10.14	7.31	10.70	4.08	−3.44	−8.69	−10.93	−7.87
34	1981	12.49	8.89	10.87	5.41	−0.97	−8.63	−13.26	−12.54
35	1982	12.17	8.29	11.31	5.32	−0.65	−6.51	−8.17	−7.33
36	1983	10.28	9.91	13.64	6.51	−2.96	−10.12	−11.67	−9.56
37	1984	11.05	13.17	10.79	5.88	−0.92	−8.21	−11.97	−9.69
38	1985	11.71	15.81	11.97	4.60	−4.02	−10.22	−12.42	−8.93
39	1986	8.26	10.17	11.07	4.20	−1.56	−6.41	−7.16	−5.81
40	1987	14.28	13.00	13.63	5.75	−0.54	−6.53	−10.94	−11.09
41	1988	12.87	13.62	12.63	5.50	−2.15	−6.79	−11.30	−10.69
42	1989	12.00	16.35	12.96	4.91	−4.55	−9.33	−11.19	−7.04
43	1990	10.92	9.85	12.16	5.77	−1.57	−8.00	−10.87	−8.85
44	1991	14.39	7.50	11.09	4.05	−2.31	−7.07	−8.61	−7.99
45	1992	13.88	11.35	11.58	4.52	−1.99	−7.24	−10.00	−8.84
46	1993	9.45	8.34	9.43	3.16	−2.57	−9.67	−13.08	−11.66
47	1994	16.84	15.64	13.51	7.50	1.58	−4.57	−8.67	−8.42

Table 7-2. Total Lake Superior basin precipitation (mm).

i	Year	$(q_{11-17Sep})_i$	$(q_{17-21Sep})_i$	$(q_{Sep})_i$	$(q_{SON})_i$	$(q_{OND})_i$	$(q_{NDJ})_i$	$(q_{DJF})_i$	$(q_{JFM})_i$
(1)	(2)	(3)	(4)	(5)	(6)	(7)	(8)	(9)	(10)
1	1948	4.20	1.00	33.00	173.81	207.92	234.60	187.20	177.30
2	1949	31.85	10.35	79.50	244.79	208.84	199.64	169.20	173.70
3	1950	15.82	14.70	54.60	227.50	229.08	186.76	150.30	177.30
4	1951	29.33	18.30	130.20	270.27	187.68	159.16	122.40	119.70
5	1952	18.90	4.90	40.50	126.49	121.44	147.20	139.50	157.50
6	1953	23.38	9.30	83.40	160.16	142.60	183.08	168.30	152.10
7	1954	8.47	30.50	91.50	200.20	131.56	107.64	116.10	170.10
8	1955	40.74	38.20	92.70	278.46	241.96	184.00	116.10	82.80
9	1956	26.67	11.30	69.60	172.90	173.88	182.16	146.70	113.40
10	1957	25.34	30.45	114.00	239.33	166.52	172.96	102.60	79.20
11	1958	12.81	8.95	83.70	227.50	196.88	181.24	108.90	87.30
12	1959	6.09	23.00	126.60	272.09	180.32	136.16	109.80	103.50
13	1960	7.70	2.70	75.60	216.58	179.40	138.00	101.70	115.20
14	1961	30.31	3.50	126.30	250.25	179.40	168.36	161.10	125.10
15	1962	14.42	8.50	91.50	154.70	119.60	120.52	121.50	107.10
16	1963	15.40	18.05	58.50	148.33	143.52	172.04	139.50	129.60
17	1964	3.78	15.15	111.00	226.59	178.48	169.28	171.00	144.00
18	1965	49.00	31.10	142.50	304.85	218.96	213.44	141.30	154.80
19	1966	14.42	3.15	49.50	209.30	216.20	188.60	165.60	151.20
20	1967	15.47	9.55	32.70	182.00	194.12	128.80	109.80	122.40
21	1968	4.55	15.70	105.60	241.15	220.80	207.00	192.60	124.20
22	1969	13.51	0.80	70.50	195.65	176.64	148.12	134.10	108.90
23	1970	16.80	21.70	118.50	312.13	253.92	195.04	187.20	175.50
24	1971	8.68	10.25	93.00	286.65	254.84	219.88	181.80	187.20
25	1972	19.60	31.35	97.20	192.92	161.92	157.32	136.80	124.20
26	1973	6.09	11.90	85.50	201.11	168.36	169.28	142.20	124.20
27	1974	25.34	9.50	78.30	217.49	174.80	200.56	163.80	177.30
28	1975	8.89	17.70	79.20	242.06	213.44	229.08	173.70	221.40
29	1976	12.11	13.15	42.00	122.85	139.84	151.80	154.80	196.20
30	1977	6.72	27.85	141.30	283.01	213.44	193.20	135.00	92.70
31	1978	33.53	3.50	74.10	191.10	178.48	180.32	157.50	197.10
32	1979	19.74	5.90	71.10	244.79	216.20	186.76	157.50	149.40
33	1980	37.66	38.45	138.90	241.15	163.76	131.56	166.50	145.80
34	1981	6.51	1.70	52.20	177.45	193.20	186.76	184.50	156.60
35	1982	37.80	17.45	108.00	300.30	266.80	192.28	150.30	121.50
36	1983	15.05	20.90	124.80	317.59	279.68	230.00	170.10	127.80
37	1984	16.03	1.40	97.80	220.22	213.44	183.08	187.20	140.40
38	1985	14.07	23.70	153.00	349.44	265.88	230.00	139.50	121.50
39	1986	22.40	18.25	96.90	237.51	171.12	128.80	83.70	77.40
40	1987	18.83	23.95	76.80	209.30	186.76	174.80	160.20	166.50
41	1988	15.19	32.40	85.20	287.56	276.92	256.68	189.00	158.40
42	1989	6.09	20.05	56.70	182.00	186.76	181.24	151.20	127.80
43	1990	42.77	22.60	95.70	251.16	206.08	148.12	122.40	128.70
44	1991	34.86	15.55	121.80	321.23	245.64	189.52	121.50	98.10
45	1992	33.04	18.05	125.40	242.06	195.04	189.52	137.70	78.30
46	1993	29.12	1.65	92.10	218.40	161.92	138.92	104.40	105.30
47	1994	29.75	6.85	76.20	200.20	141.68	126.04	109.80	126.00

Table 7-3. Ordered average Lake Superior basin air temperatures 1961–90 (°C).

i	i/n	$(t_{11-17\text{Sep}})_i$	$(t_{17-21\text{Sep}})_i$	$(t_{\text{Sep}})_i$	$(t_{\text{SON}})_i$	$(t_{\text{OND}})_i$	$(t_{\text{NDJ}})_i$	$(t_{\text{DJF}})_i$	$(t_{\text{JFM}})_i$
(1)	(2)	(3)	(4)	(5)	(6)	(7)	(8)	(9)	(10)
1	0.033	8.26	6.99	9.14	3.51	−5.36	−12.28	−15.22	−13.31
2	0.067	8.59	7.31	9.44	4.02	−4.55	−10.79	−14.22	−13.01
3	0.100	8.67	7.96	10.24	4.08	−4.02	−10.22	−14.12	−12.68
4	0.133	8.77	8.18	10.37	4.20	−3.63	−10.12	−13.26	−12.54
5	0.167	9.72	8.29	10.70	4.52	−3.44	−9.33	−13.16	−12.06
6	0.200	9.79	8.89	10.79	4.60	−2.96	−9.12	−13.09	−11.95
7	0.233	10.14	9.42	10.87	4.61	−2.95	−9.10	−12.92	−11.94
8	0.267	10.19	9.85	10.93	4.67	−2.64	−8.97	−12.71	−11.45
9	0.300	10.28	9.91	11.04	4.74	−2.46	−8.84	−12.54	−11.42
10	0.333	10.38	10.17	11.07	4.91	−2.15	−8.79	−12.49	−11.15
11	0.367	10.78	10.30	11.13	5.04	−1.69	−8.69	−12.49	−11.09
12	0.400	10.92	10.67	11.31	5.12	−1.57	−8.63	−12.42	−10.99
13	0.433	11.05	10.81	11.58	5.15	−1.56	−8.56	−12.10	−10.87
14	0.467	11.45	10.92	11.83	5.20	−1.44	−8.56	−11.97	−10.69
15	0.500	11.69	11.24	11.86	5.32	−1.36	−8.49	−11.67	−10.01
16	0.533	11.71	11.40	11.97	5.41	−1.36	−8.45	−11.47	−9.87
17	0.567	11.92	11.51	12.05	5.50	−1.21	−8.26	−11.30	−9.69
18	0.600	12.00	11.69	12.16	5.75	−1.19	−8.21	−11.19	−9.69
19	0.633	12.15	12.22	12.19	5.75	−1.02	−8.00	−11.18	−9.57
20	0.667	12.17	12.65	12.29	5.77	−0.98	−8.00	−11.17	−9.56
21	0.700	12.49	12.87	12.31	5.84	−0.97	−7.85	−10.94	−9.56
22	0.733	12.49	13.00	12.43	5.88	−0.92	−7.74	−10.93	−9.46
23	0.767	12.68	13.17	12.57	6.16	−0.85	−7.54	−10.87	−8.93
24	0.800	12.87	13.62	12.63	6.27	−0.79	−6.96	−10.51	−8.85
25	0.833	13.17	14.87	12.96	6.47	−0.65	−6.79	−10.40	−8.58
26	0.867	13.48	15.43	13.05	6.51	−0.54	−6.53	−10.31	−7.87
27	0.900	13.51	15.75	13.37	6.55	−0.51	−6.51	−10.21	−7.33
28	0.933	14.28	15.81	13.45	6.63	−0.34	−6.44	−9.16	−7.33
29	0.967	15.50	15.95	13.63	7.04	−0.32	−6.41	−8.17	−7.04
30	1.000	15.88	16.35	13.64	8.40	0.30	−6.08	−7.16	−5.81

Readers may find these equations for themselves by following the directions in Appendix 2 for Exercise A2-4. Equation (7-15) must be solved to find the weights, but the number of equations represented there is already less than the number of weights and may be reduced further if any equations prove redundant or infeasible. Therefore, multiple solutions exist, and the optimization of Equations (7-5) is used to find the "best."

The Lagrangian of Equation (7-6), for the optimization of Equations (7-5) associated with this problem is differentiated with respect to all its variables, and those differentials are set to zero, as in Equations (7-7) and (7-8). Table 7-5 presents the solution of Equation (7-15), found by using the *first* procedural algorithm in Figure 7-2 (using all scenarios). A check reveals that $\sum w_i^2 < 2n$; therefore, the solution is a minimum. Table 7-6 presents the solution of Equation (7-15) found by using the *second* procedural algorithm in Figure 7-2 (maximizing the number of meteorology event probabilities used). Again, a check reveals that $\sum w_i^2 < 2n$, so the solution is a minimum. All computa-

Table 7-4. Ordered total Lake Superior basin precipitation 1961–90 (mm).

i	i/n	$\left(q_{11-17\text{Sep}}\right)_i$	$\left(q_{17-21\text{Sep}}\right)_i$	$\left(q_{\text{Sep}}\right)_i$	$\left(q_{\text{SON}}\right)_i$	$\left(q_{\text{OND}}\right)_i$	$\left(q_{\text{NDJ}}\right)_i$	$\left(q_{\text{DJF}}\right)_i$	$\left(q_{\text{JFM}}\right)_i$
(1)	(2)	(3)	(4)	(5)	(6)	(7)	(8)	(9)	(10)
1	0.033	3.78	0.80	32.70	122.85	119.60	120.52	83.70	77.40
2	0.067	4.55	1.40	42.00	148.33	139.84	128.80	109.80	92.70
3	0.100	6.09	1.70	49.50	154.70	143.52	128.80	121.50	107.10
4	0.133	6.09	3.15	52.20	177.45	161.92	131.56	122.40	108.90
5	0.167	6.51	3.50	56.70	182.00	163.76	148.12	134.10	121.50
6	0.200	6.72	3.50	58.50	182.00	168.36	148.12	135.00	121.50
7	0.233	8.68	5.90	70.50	191.10	171.12	151.80	136.80	122.40
8	0.267	8.89	8.50	71.10	192.92	174.80	157.32	139.50	124.20
9	0.300	12.11	9.50	74.10	195.65	176.64	168.36	139.50	124.20
10	0.333	13.51	9.55	76.80	201.11	178.48	169.28	141.30	124.20
11	0.367	14.07	10.25	78.30	209.30	178.48	169.28	142.20	125.10
12	0.400	14.42	11.90	79.20	209.30	179.40	172.04	150.30	127.80
13	0.433	14.42	13.15	85.20	217.49	186.76	174.80	151.20	127.80
14	0.467	15.05	15.15	85.50	220.22	186.76	180.32	154.80	128.70
15	0.500	15.19	15.70	91.50	226.59	193.20	181.24	157.50	129.60
16	0.533	15.40	17.45	93.00	237.51	194.12	183.08	157.50	140.40
17	0.567	15.47	17.70	95.70	241.15	206.08	186.76	160.20	144.00
18	0.600	16.03	18.05	96.90	241.15	213.44	186.76	161.10	145.80
19	0.633	16.80	18.25	97.20	242.06	213.44	188.60	163.80	149.40
20	0.667	18.83	20.05	97.80	244.79	213.44	192.28	165.60	151.20
21	0.700	19.60	20.90	105.60	250.25	216.20	193.20	166.50	154.80
22	0.733	19.74	21.70	108.00	251.16	216.20	195.04	170.10	156.60
23	0.767	22.40	22.60	111.00	283.01	218.96	200.56	171.00	158.40
24	0.800	25.34	23.70	118.50	286.65	220.80	207.00	173.70	166.50
25	0.833	30.31	23.95	124.80	287.56	253.92	213.44	181.80	175.50
26	0.867	33.53	27.85	126.30	300.30	254.84	219.88	184.50	177.30
27	0.900	37.66	31.10	138.90	304.85	265.88	229.08	187.20	187.20
28	0.933	37.80	31.35	141.30	312.13	266.80	230.00	187.20	196.20
29	0.967	42.77	32.40	142.50	317.59	276.92	230.00	189.00	197.10
30	1.000	49.00	38.45	153.00	349.44	279.68	256.68	192.60	221.40

tions were made with probabilities (both reference quantiles and forecasts) significant to three digits after the decimal point.

Weights from the first procedural algorithm (using all meteorology time series segments) are given in Table 7-5. They may be computed as in Exercise A2-5 in Appendix 2. While all 47 scenarios are used (all weights are strictly positive), not all of the selected meteorology event probabilities can be used. All of the temperature event and precipitation event probabilities for September 11–17, 1998, were used, while the temperature event probabilities for SON [Equations (7-13e) and (7-13f) or the last two rows of Equation (7-15)] were unused. Weights from the second procedural algorithm (maximizing use of meteorology outlooks) are given in Table 7-6 (also see Exercise A2-5 in Appendix 2). Now, all 47 scenarios are not used (there are zero weights indicated for indices 6, 9, and 14), but all of the selected meteorology event probabilities are used [all Equations (7-13) or Equation (7-15)].

Table 7-5. Outlook weights using all meteorology time series segments for Equation (7-15).

Index, i (1)	Weight, w_i (2)	Index, i (3)	Weight, w_i (4)	Index, i (5)	Weight, w_i (6)
1	1.498014	17	0.768555	33	0.123474
2	0.978780	18	0.123474	34	1.498014
3	1.651245	19	1.525398	35	0.978780
4	0.852933	20	1.525398	36	0.795939
5	0.852933	21	1.498014	37	1.651245
6	0.123474	22	1.498014	38	1.651245
7	0.768555	23	0.795939	39	0.123474
8	0.852933	24	1.623861	40	1.525398
9	0.123474	25	0.978780	41	1.525398
10	0.852933	26	0.768555	42	1.623861
11	1.498014	27	0.123474	43	0.978780
12	0.768555	28	0.768555	44	0.852933
13	0.768555	29	1.498014	45	0.852933
14	0.123474	30	1.623861	46	0.123474
15	1.525398	31	0.978780	47	0.852933
16	1.525398	32	0.978780		

The weights may now be applied to derivative future scenarios corresponding to the segments of the historical meteorology record. Again, derivative hydrology scenarios, beginning September 5, 1998, for the Lake Superior basin, are constructed with the hypothetical snowmelt model of Equations (5-1) and (5-2) and Figure 5-2, with no initial snow pack. This is similar to the example in Chapter 6, except the Lake Superior data set is used. The first 5 days of each segment of the historical meteorology record beginning on September 5 on the Lake Superior basin are shown in Table 7-7. Derivative hy-

Table 7-6. Outlook weights maximizing use of meteorology outlooks for Equation (7-15).

Index, i (1)	Weight, w_i (2)	Index, i (3)	Weight, w_i (4)	Index, i (5)	Weight, w_i (6)
1	1.278181	17	0.775781	33	0.372451
2	0.706473	18	0.372451	34	1.567911
3	1.678484	19	2.136591	35	0.706473
4	1.164581	20	1.570379	36	0.488519
5	0.598368	21	1.278181	37	1.388754
6	0	22	1.567911	38	2.244696
7	0.486051	23	0.488519	39	0.372451
8	0.598368	24	1.386286	40	1.570379
9	0	25	1.272686	41	1.570379
10	0.598368	26	0.486051	42	2.242229
11	1.278181	27	0.372451	43	0.706473
12	1.341993	28	0.775781	44	1.164581
13	0.486051	29	2.134123	45	1.164581
14	0	30	1.386286	46	0.372451
15	1.280649	31	0.706473	47	0.308638
16	1.280649	32	1.272686		

Table 7-7. Areal average Lake Superior basin air temperatures (T) and precipitation (Q).

Year (1)	Type (2)	5 Sep (3)	6 Sep (4)	7 Sep (5)	8 Sep (6)	9 Sep (7)	Year (8)	5 Sep (9)	6 Sep (10)	7 Sep (11)	8 Sep (12)	9 Sep (13)
1948	T (°C)	19.96	19.49	16.90	13.13	11.89	1972	10.87	11.39	12.93	10.36	10.46
	Q (mm)	2.22	9.60	6.07	1.12	3.13		1.42	11.66	2.49	0.07	0.05
1949	T (°C)	10.86	10.13	9.92	9.27	12.24	1973	16.95	14.02	11.76	10.82	11.75
	Q (mm)	2.73	1.37	2.01	1.51	0.04		1.17	0.25	0.00	0.00	0.00
1950	T (°C)	11.86	15.53	15.48	15.26	12.65	1974	12.99	12.95	13.59	13.51	10.36
	Q (mm)	0.00	0.00	0.00	0.00	0.00		0.00	0.07	0.66	0.67	4.42
1951	T (°C)	11.94	11.95	9.08	13.77	14.81	1975	12.73	10.71	9.98	8.28	7.77
	Q (mm)	4.22	6.57	0.29	0.07	5.71		10.52	4.13	17.31	1.41	0.37
1952	T (°C)	15.04	9.34	8.54	15.11	17.64	1976	10.83	15.89	19.28	19.72	14.45
	Q (mm)	1.61	1.56	0.09	0.20	1.33		0.01	0.03	0.06	1.35	1.67
1953	T (°C)	13.38	11.94	11.36	12.04	12.88	1977	12.26	11.72	10.96	12.01	14.35
	Q (mm)	1.26	0.65	0.47	0.33	0.97		0.79	4.02	0.50	27.74	16.85
1954	T (°C)	14.91	11.44	12.39	10.41	9.17	1978	17.62	18.65	14.39	13.43	13.20
	Q (mm)	0.24	9.67	1.61	0.15	8.48		2.25	4.68	2.44	0.19	0.79
1955	T (°C)	14.57	13.62	9.10	10.48	12.08	1979	19.33	14.71	9.64	9.17	12.68
	Q (mm)	0.44	0.05	0.09	2.35	6.40		2.83	1.89	0.93	0.43	1.41
1956	T (°C)	10.77	9.69	7.48	7.25	8.57	1980	16.07	14.52	16.41	17.07	14.85
	Q (mm)	9.85	4.34	1.78	0.38	4.75		0.08	0.49	0.26	11.60	6.98
1957	T (°C)	10.63	11.30	10.52	12.16	14.54	1981	11.60	12.88	14.46	12.17	12.25
	Q (mm)	0.03	0.04	0.00	0.00	0.51		0.01	1.32	10.95	2.22	1.60
1958	T (°C)	11.98	12.37	12.44	12.16	12.74	1982	12.37	9.15	9.30	13.73	17.66
	Q (mm)	6.87	8.69	5.22	0.27	3.50		3.30	0.56	0.69	0.49	0.15
1959	T (°C)	18.45	20.39	19.72	17.17	19.78	1983	19.05	16.60	14.80	17.29	20.02
	Q (mm)	1.81	23.51	5.77	0.57	9.75		12.53	6.29	0.62	21.33	4.21
1960	T (°C)	17.51	21.62	20.69	16.08	13.74	1984	9.91	11.91	15.56	16.01	13.82
	Q (mm)	0.60	4.22	2.95	4.89	0.66		0.56	0.54	2.11	8.54	4.07
1961	T (°C)	14.06	16.23	16.71	18.59	21.37	1985	16.08	17.82	16.01	14.35	12.43
	Q (mm)	3.91	9.54	0.87	2.96	3.94		0.53	0.78	0.70	1.58	1.14
1962	T (°C)	9.10	11.95	14.85	16.54	15.19	1986	10.86	6.72	7.54	11.13	10.33
	Q (mm)	1.28	0.03	0.21	4.70	9.25		2.04	0.69	0.74	0.16	1.86
1963	T (°C)	14.79	15.91	12.85	14.97	13.90	1987	17.65	17.19	15.20	15.05	14.72
	Q (mm)	0.18	6.46	1.75	1.57	0.86		8.23	6.01	0.90	3.50	0.93
1964	T (°C)	12.93	13.88	14.55	13.20	13.86	1988	9.79	11.37	13.45	15.52	14.51
	Q (mm)	0.34	6.01	14.45	3.38	9.24		1.42	0.59	0.00	2.42	0.13
1965	T (°C)	12.31	7.59	10.66	11.53	14.10	1989	17.47	17.92	18.81	16.98	14.67
	Q (mm)	1.85	4.53	3.15	2.68	4.24		4.23	0.17	1.89	1.23	1.33
1966	T (°C)	13.49	13.81	13.61	16.43	17.82	1990	15.98	18.00	13.34	11.74	16.02
	Q (mm)	3.06	0.21	0.00	0.00	0.86		2.72	8.25	0.55	0.08	1.18
1967	T (°C)	17.16	17.72	17.48	17.72	9.76	1991	12.30	11.60	15.67	16.50	18.85
	Q (mm)	0.00	0.00	0.00	0.24	0.03		1.58	0.51	3.62	8.53	12.38
1968	T (°C)	13.63	11.21	9.04	11.34	12.51	1992	17.66	15.04	11.15	11.81	10.39
	Q (mm)	9.53	5.83	4.85	8.31	11.23		4.65	4.12	8.40	5.39	8.22
1969	T (°C)	18.63	17.60	12.93	10.88	9.42	1993	9.41	10.14	10.09	9.88	12.19
	Q (mm)	7.74	4.94	2.20	1.08	0.31		1.37	3.10	1.08	5.49	11.41
1970	T (°C)	15.76	17.89	20.18	16.54	16.43	1994	14.49	14.56	14.64	13.52	11.80
	Q (mm)	0.01	0.53	5.33	0.65	12.14		5.15	1.57	0.56	0.05	0.00
1971	T (°C)	19.97	18.92	17.49	17.25	14.28						
	Q (mm)	4.21	2.27	1.96	0.62	4.50						

Table 7-8. Lake Superior basin runoff (mm) transformed from Table 7-7 with Equations (5-1) and (5-2).

Index, i	5 Sep	6 Sep	7 Sep	8 Sep	9 Sep	Index, i	5 Sep	6 Sep	7 Sep	8 Sep	9 Sep
(1)	(2)	(3)	(4)	(5)	(6)	(7)	(8)	(9)	(10)	(11)	(12)
1	2.22	9.60	6.07	1.12	3.13	25	1.42	11.66	2.49	0.07	0.05
2	2.73	1.37	2.01	1.51	0.04	26	1.17	0.25	0.00	0.00	0.00
3	0.00	0.00	0.00	0.00	0.00	27	0.00	0.07	0.66	0.67	4.42
4	4.22	6.57	0.29	0.07	5.71	28	10.52	4.13	17.31	1.41	0.37
5	1.61	1.56	0.09	0.20	1.33	29	0.01	0.03	0.06	1.35	1.67
6	1.26	0.65	0.47	0.33	0.97	30	0.79	4.02	0.50	27.74	16.85
7	0.24	9.67	1.61	0.15	8.48	31	2.25	4.68	2.44	0.19	0.79
8	0.44	0.05	0.09	2.35	6.40	32	2.83	1.89	0.93	0.43	1.41
9	9.85	4.34	1.78	0.38	4.75	33	0.08	0.49	0.26	11.60	6.98
10	0.03	0.04	0.00	0.00	0.51	34	0.01	1.32	10.95	2.22	1.60
11	6.87	8.69	5.22	0.27	3.50	35	3.30	0.56	0.69	0.49	0.15
12	1.81	23.51	5.77	0.57	9.75	36	12.53	6.29	0.62	21.33	4.21
13	0.60	4.22	2.95	4.89	0.66	37	0.56	0.54	2.11	8.54	4.07
14	3.91	9.54	0.87	2.96	3.94	38	0.53	0.78	0.70	1.58	1.14
15	1.28	0.03	0.21	4.70	9.25	39	2.04	0.69	0.74	0.16	1.86
16	0.18	6.46	1.75	1.57	0.86	40	8.23	6.01	0.90	3.50	0.93
17	0.34	6.01	14.45	3.38	9.24	41	1.42	0.59	0.00	2.42	0.13
18	1.85	4.53	3.15	2.68	4.24	42	4.23	0.17	1.89	1.23	1.33
19	3.06	0.21	0.00	0.00	0.86	43	2.72	8.25	0.55	0.08	1.18
20	0.00	0.00	0.00	0.24	0.03	44	1.58	0.51	3.62	8.53	12.38
21	9.53	5.83	4.85	8.31	11.23	45	4.65	4.12	8.40	5.39	8.22
22	7.74	4.94	2.20	1.08	0.31	46	1.37	3.10	1.08	5.49	11.41
23	0.01	0.53	5.33	0.65	12.14	47	5.15	1.57	0.56	0.05	0.00
24	4.21	2.27	1.96	0.62	4.50						

drology is given in Table 7-8 and weighted by the values in Table 7-5 (using all meteorology time series segments) as in Equation (5-11a) to yield probabilistic hydrology outlook estimates in Table 7-9. To illustrate, consider columns 3 and 9 in Table 7-8 for runoff on September 6, 1998. Table 7-10 details the computations used in applying Equation (5-11a) to make this forecast. In Table 7-10, column 2 is taken from columns 3 and 9 in Table 7-8 and column 3 is taken from Table 7-5. Column 2 in Table 7-10 is ordered from smallest to largest in column 4; columns 5 and 6 are the indices and weights corresponding to the entries in column 4, as can be seen by inspection. Column 7 is the running sum of weights in column 6, and column 8 is the running sum in column 7 divided by the number of weights ($n = 47$). Column 8 is thus the nonexceedance probability estimate of Equation (5-11a) for each value in column 4. Finally, the values in Table 7-9, column 3, are interpolated in column 9 of Table 7-10 from the distribution in column 8. From Table 7-9, for example, the forecast is for a 2/3 probability of not exceeding 4.59 mm on September 6, 1998 ($\hat{P}\left[R_{6Sep98} \leq 4.59 \text{ mm}\right] = 0.667$ or $\hat{r}_{6Sep98, 0.667}$ = 4.59 mm). (Again, do not confuse these forecast quantiles with historical quantiles estimated from the historical record.) Stated another way, the forecast probability that runoff on September 6, 1998, will exceed 4.59 mm is 1/3.

Table 7-9. Hydrology (runoff) outlook for Equation (7-15) using all meteorology time series segments (mm).

Outlook variable (1)	Outlook day, j				
	5 Sep 98 (2)	6 Sep 98 (3)	7 Sep 98 (4)	8 Sep 98 (5)	9 Sep 98 (6)
$\hat{r}_{j,\,0.100}$	0.01	0.03	0.00	0.00	0.03
$\hat{r}_{j,\,0.250}$	0.41	0.24	0.14	0.21	0.41
$\hat{r}_{j,\,0.333}$	0.58	0.55	0.52	0.44	0.86
$\hat{r}_{j,\,0.500}$	1.57	1.56	1.01	1.17	1.33
$\hat{r}_{j,\,0.667}$	2.90	4.59	2.08	1.82	3.95
$\hat{r}_{j,\,0.750}$	4.21	5.91	2.49	3.16	4.49
$\hat{r}_{j,\,0.900}$	7.74	8.52	5.91	8.40	9.94
$\bar{r}$	2.80	3.48	2.52	3.20	3.59
$\sqrt{s_R^2}$	3.09	4.15	3.53	5.81	4.38

The transformed time series in Table 7-8 are weighted by the values in Table 7-6 (maximizing use of meteorology outlooks) and used in Equations (5-11) to yield the representative probabilistic hydrology outlook estimates in Table 7-11. From Table 7-11, for example, the forecast is for a 2/3 probability of not exceeding 4.20 mm on September 6, 1998. Compared with the forecast in Table 7-9, the effect of including the September-October-November precipitation outlook in the example lowered the estimated forecast 2/3 quantile of runoff on September 6, 1998. Stated another way, it lowered the estimated forecast probability of exceeding 4.59 mm on September 6, 1998. (The estimated probability of not exceeding 4.59 mm is 2/3 in Table 7-9 and more than 2/3 in Table 7-11, by interpolation; therefore, the estimated probability of exceeding 4.59 mm is 1/3 in Table 7-9 and less than 1/3 in Table 7-11.)

Lake Superior Supply Outlook
As a second example, consider the following. NOAA's Great Lakes Environmental Research Laboratory (GLERL) has an integrated system of computer models for basin moisture, rainfall runoff, evapotranspiration, over-lake precipitation, lake thermodynamics (including heat storage, water surface temperature, and lake evaporation), connecting channel hydraulics, and lake regulation (Croley et al. 1996) as part of its *Advanced Hydrologic Prediction System* (Croley 1998). Inputs are daily minimum, maximum, and average air temperatures, daily precipitation, daily humidity, daily wind speed, and daily cloud cover, for each of the watersheds and the lake surface comprising a Great Lake basin. GLERL used its hydrology models to estimate the 12.5-month probabilistic outlook of net basin supply and other hydrology variables for Lake Superior beginning September 15, 1998, by using the first 24 probabilities from the NOAA Climate Prediction Center *Climate Outlook* for September 1998. (Net basin supply is the algebraic sum of over-lake precipitation, lake evaporation, and basin runoff to the lake.) Recall that this meteorology outlook was presented in the example of Figures 4-3 and 4-4, with probabilities abstracted for the Lake Superior basin and presented in Figure 4-5. Note that the first eight probability statements to be used cover periods starting September 1 but the derivative outlook begins September 15. This is another example of using a meteorology outlook that begins earlier than the derivative outlook. Per the dis-

Table 7-10. Computation of hydrology (runoff) outlook quantiles from Tables 7-5 and 7-8.

	Unordered (original)		Ordered					Interpolations
i	$(r_{6Sep98})_i$	w_i	$(r_{6Sep98})_i$	i	w_i	$\sum w_i$	$\frac{1}{n}\sum w_i$	
(1)	(2)	(3)	(4)	(5)	(6)	(7)	(8)	(9)
1	9.60	1.498014	0.00	3	1.651245	1.651245	0.035133	
2	1.37	0.978780	0.00	20	1.525398	3.176643	0.067588	
3	0.00	1.651245	0.03	29	1.498014	4.674657	0.099461	@ 0.100, $\hat{r}_{6Sep98,\,0.100} = 0.03$
4	6.57	0.852933	0.03	15	1.525398	6.200055	0.131916	
5	1.56	0.852933	0.04	10	0.852933	7.052988	0.150064	
6	0.65	0.123474	0.05	8	0.852933	7.905921	0.168211	
7	9.67	0.768555	0.07	27	0.123474	8.029395	0.170838	
8	0.05	0.852933	0.17	42	1.623861	9.653256	0.205388	
9	4.34	0.123474	0.21	19	1.525398	11.178650	0.237844	@ 0.250, $\hat{r}_{6Sep98,\,0.250} = 0.24$
10	0.04	0.852933	0.25	26	0.768555	11.947210	0.254196	
11	8.69	1.498014	0.49	33	0.123474	12.070680	0.256823	
12	23.51	0.768555	0.51	44	0.852933	12.923620	0.274971	
13	4.22	0.768555	0.53	23	0.795939	13.719560	0.291905	
14	9.54	0.123474	0.54	37	1.651245	15.370800	0.327038	@ 0.333, $\hat{r}_{6Sep98,\,0.333} = 0.55$
15	0.03	1.525398	0.56	35	0.978780	16.349580	0.347863	
16	6.46	1.525398	0.59	41	1.525398	17.874980	0.380319	
17	6.01	0.768555	0.65	6	0.123474	17.998450	0.382946	
18	4.53	0.123474	0.69	39	0.123474	18.121930	0.385573	
19	0.21	1.525398	0.78	38	1.651245	19.773170	0.420706	
20	0.00	1.525398	1.32	34	1.498014	21.271190	0.452578	
21	5.83	1.498014	1.37	2	0.978780	22.249970	0.473404	
22	4.94	1.498014	1.56	5	0.852933	23.102900	0.491551	@ 0.500, $\hat{r}_{6Sep98,\,0.500} = 1.56$
23	0.53	0.795939	1.57	47	0.852933	23.955840	0.509699	
24	2.27	1.623861	1.89	32	0.978780	24.934620	0.530524	

Continued on next page.

Table 7-10 (continued). Computation of hydrology (runoff) outlook quantiles from Tables 7-5 and 7-8.

	Unordered (original)			Ordered				Interpolations
i	$(r_{6Sep98})_i$	w_i	$(r_{6Sep98})_i$	i	w_i	$\sum w_i$	$\frac{1}{n}\sum w_i$	
(1)	(2)	(3)	(4)	(5)	(6)	(7)	(8)	(9)
25	11.66	0.978780	2.27	24	1.623861	26.558480	0.565074	
26	0.25	0.768555	3.10	46	0.123474	26.681950	0.567701	
27	0.07	0.123474	4.02	30	1.623861	28.305810	0.602251	
28	4.13	0.768555	4.12	45	0.852933	29.158750	0.620399	
29	0.03	1.498014	4.13	28	0.768555	29.927300	0.636751	
30	4.02	1.623861	4.22	13	0.768555	30.695860	0.653103	
31	4.68	0.978780	4.34	9	0.123474	30.819330	0.655730	
32	1.89	0.978780	4.53	18	0.123474	30.942800	0.658358	@ 0.667, $\hat{r}_{6Sen98,\,0.667}$ = 4.59
33	0.49	0.123474	4.68	31	0.978780	31.921590	0.679183	
34	1.32	1.498014	4.94	22	1.498014	33.419600	0.711055	
35	0.56	0.978780	5.83	21	1.498014	34.917610	0.742928	
36	6.29	0.795939	6.01	17	0.768555	35.686160	0.759280	@ 0.750, $\hat{r}_{6Sen98,\,0.750}$ = 5.91
37	0.54	1.651245	6.01	40	1.525398	37.211560	0.791735	
38	0.78	1.651245	6.29	36	0.795939	38.007500	0.808670	
39	0.69	0.123474	6.46	16	1.525398	39.532900	0.841126	
40	6.01	1.525398	6.57	4	0.852933	40.385830	0.859273	
41	0.59	1.525398	8.25	43	0.978780	41.364610	0.880098	@ 0.900, $\hat{r}_{6Sen98,\,0.900}$ = 8.52
42	0.17	1.623861	8.69	11	1.498014	42.862630	0.911971	
43	8.25	0.978780	9.54	14	0.123474	42.986100	0.914598	
44	0.51	0.852933	9.60	1	1.498014	44.484110	0.946470	
45	4.12	0.852933	9.67	7	0.768555	45.252670	0.962823	
46	3.10	0.123474	11.66	25	0.978780	46.231450	0.983648	
47	1.57	0.852933	23.51	12	0.768555	47.000000	1.000000	

Table 7-11. Hydrology (runoff) outlook for Equation (7-15) maximizing use of meteorology outlooks (mm).

Outlook variable (1)	Outlook day, j				
	5 Sep 98 (2)	6 Sep 98 (3)	7 Sep 98 (4)	8 Sep 98 (5)	9 Sep 98 (6)
$\hat{r}_{j,\,0.100}$	0.01	0.02	0.00	0.00	0.04
$\hat{r}_{j,\,0.250}$	0.41	0.19	0.13	0.23	0.66
$\hat{r}_{j,\,0.333}$	0.57	0.53	0.51	0.51	0.86
$\hat{r}_{j,\,0.500}$	1.57	1.36	0.93	1.20	1.35
$\hat{r}_{j,\,0.667}$	2.87	4.20	2.10	1.65	3.96
$\hat{r}_{j,\,0.750}$	4.21	5.76	2.70	2.67	4.85
$\hat{r}_{j,\,0.900}$	7.41	8.58	5.97	8.33	9.75
$\bar{r}$	2.68	3.55	2.56	3.00	3.60
$\sqrt{s_R^2}$	2.92	4.69	3.59	5.36	4.30

cussion in Chapter 6, these first eight equations (with periods not fully contained in the 12.5-month forecast period starting September 15) will be actually imperfectly satisfied.

The outlook begins by identifying all 12.5-month meteorology time series starting September 15 from the available historical record of 1948–94; there are 47 such time series for each meteorology variable. These time series, for all meteorology variables, are used in simulations with GLERL's hydrology models and current (September 15, 1998) initial conditions to estimate the 47 associated time series for each *hydrology* variable. Table 7-12 contains the transformed hydrology time series segments for monthly net basin supply corresponding to each meteorology time series segment taken from the historical record. Now, make a probabilistic hydrology outlook by using the first 24 probabilistic meteorology outlooks presented in the example of Figure 4-5, prioritized by their order of listing there (see Exercise A2-6 in Appendix 2). By using the average air temperature and total precipitation over the Lake Superior basin for the periods of September, SON, OND, NDJ, DJF, and JFM in the historical record from 1961 to 1990 (in Tables 7-1 and 7-2), the relevant reference quantile estimates can be determined from the ordered values in Tables 7-3 and 7-4. The first 24 multiple meteorology event probabilities of Figure 4-5 are expressed, after inserting Equation (6-3) as the highest-priority equation, as the vector equation in Figure 7-3. The component equations are derived and can be checked, as before, by noting the inclusion or exclusion of each value of monthly or 3-monthly air temperature or precipitation in the respective sets indicated in the variable subscripts in Figure 4-5. For example, the first equation in Figure 4-5 requires the set of all historical meteorology time series for which the September air temperature (column 5 in Table 7-1) is less than or equal to the 1/3 quantile (column 5 in Table 7-3 for index 10 gives 11.07°C). By looking at each entry in column 5 of Table 7-1 and placing a 1 if it is less than or equal to 11.07°C and a 0 if it is greater than 11.07°C, the second line in the matrix in Figure 7-3 is derived.

The Lagrangian of Equation (7-6), for the optimization of Equations (7-5) that is associated with this problem is differentiated with respect to all its variables and those differentials are set to zero, as in Equations (7-7) and (7-8). Table 7-13 presents the solution of the equations represented in Figure 7-3 found by using the second procedural algorithm in Figure 7-2 (maximizing use of meteorology outlooks). Checks reveal that $\sum w_i^2 < 2n$ for both procedural algorithms; therefore the solutions represent minimums. (All computations are with probabilities, both reference quantiles and forecasts, significant to three digits after the decimal point.)

The first procedural algorithm (using all meteorology time series segments; weights not shown) matches the first 13 equations in Figure 4-5 while using all of the meteorology time series segments from 1948–94. The second procedural algorithm (maximizing use of meteorology outlooks) matches the first 24 equations in Figure 4-5 (as intended) but has zero weights for years 1950, 1951, 1952, 1955, 1956, 1958, 1960, 1962, 1963, 1986, 1991, and 1994; see Table 7-13. By using this set of weights, the probabilistic hydrology outlooks for Lake Superior net basin supply, shown in Table 7-14, are built from Table 7-12. See Exercise A2-6 in Appendix 2 for this example, worked with the software described there.

El Niño and La Niña Example

Consider the La Niña meteorology forecasts of Equations (4-9), prioritized by their order of listing there, for another 12.5-month outlook of Lake Superior net basin supply starting September 15, 1998. This is an example of using meteorology outlooks that begin in the derivative outlook period. Per the discussion in Chapter 6, these outlooks all begin after September 15 and may be *perfectly* satisfied with actual as well as historical data, if only they form a feasible set of equations for the data set used here. Again, Table 7-12 contains the transformed hydrology time series segments for net basin supply corresponding to each meteorology time series segment taken from the historical record. The multiple meteorology event probabilities of Equations (4-9) are expressed, after inserting Equation (6-3) as the highest-priority equation (see Exercise A2-7 in Appendix 2), as

$$
\begin{bmatrix}
11111111111111111111111111111111111111111111111 \\
00110001001100001010000010001010100101001101010 \\
10001110110001110100000100100000011010010000001 \\
01110000101000101000001101001010000101000100010 \\
10001110010100010101100000100001001000111001101 \\
00000000101001101010000101001110010000000000010 \\
10011111010100010100100000100001101000110011101 \\
01000000001000101010011101000110010000000000010 \\
00011101010010010000000010101000101001100111101
\end{bmatrix}
\begin{bmatrix} w_1 \\ w_2 \\ \vdots \\ w_{47} \end{bmatrix}
=
\begin{bmatrix}
1.000 \times 47 \\
0.412 \times 47 \\
0.235 \times 47 \\
0.588 \times 47 \\
0.176 \times 47 \\
0.529 \times 47 \\
0.235 \times 47 \\
0.529 \times 47 \\
0.176 \times 47
\end{bmatrix}
$$

(7-16)

The component equations are derived, as before, by noting the inclusion or exclusion of each value of 3-monthly air temperature in the respective sets indicated in the variable subscripts in Equations (4-9). For example, Equation (4-9a) requires the set of all historical meteorology time series for which the October-November-December air temperature (column 7 in Table 7-1) is less than or equal to the 1/3 quantile (column 7 in Table 7-3 for index 10 gives $-2.15°C$). By looking at each entry in column 7 of Table 7-1 and inserting a 1 if it is less than or equal to $-2.15°C$ and a 0 if it is greater than $-2.15°C$, derive the second line in Equation (7-16).

The Lagrangian of Equation (7-6) for the optimization of Equations (7-5) that is associated with Equation (7-16) is differentiated with respect to all its variables and those

95

Table 7-12. Lake Superior net basin supply (mm) transformed from 1948–94 historical meteorology with September 15, 1998, initial conditions.

Index, i (1)	Sep 98 (2)	Oct 98 (3)	Nov 98 (4)	Dec 98 (5)	Jan 99 (6)	Feb 99 (7)	Mar 99 (8)	Apr 99 (9)	May 99 (10)	Jun 99 (11)	Jul 99 (12)	Aug 99 (13)	Sep 99 (14)
1	-2.02	-26.40	34.89	-49.54	-36.37	-28.83	8.66	71.71	162.00	164.46	134.13	53.67	31.33
2	-3.37	67.92	-17.60	-74.29	-39.00	-39.03	-4.31	71.51	246.09	201.70	148.55	83.37	51.85
3	19.01	53.58	-5.19	-56.71	-67.03	8.35	33.84	136.97	143.26	176.86	104.58	113.54	123.83
4	42.80	52.74	-20.26	-40.06	-33.98	-21.62	1.65	115.69	118.06	170.43	193.72	98.61	26.16
5	5.53	-85.05	-13.12	-37.68	-32.73	-16.24	37.42	98.71	192.75	198.35	125.03	111.51	33.65
6	24.12	-15.23	-12.24	-28.90	-51.08	4.70	-7.44	132.57	211.65	155.20	78.96	66.55	73.11
7	51.84	28.17	-12.66	-51.87	-53.10	-19.20	-5.09	140.68	133.48	101.30	117.42	80.95	45.45
8	72.02	64.68	-4.91	-55.87	-25.27	-41.41	-29.58	78.40	170.51	146.14	121.96	81.96	55.77
9	3.31	-11.84	-20.03	-36.89	-80.69	-24.25	3.28	115.55	152.77	168.10	118.98	41.02	110.26
10	43.19	-14.79	17.86	-45.16	-25.07	-51.41	11.06	76.07	96.71	147.16	138.31	94.70	88.52
11	28.23	3.45	17.66	-60.48	-68.54	-37.59	1.93	69.30	196.96	127.23	104.79	158.48	103.94
12	62.46	23.93	-48.84	-43.31	-34.06	-33.54	-28.17	161.70	202.50	134.05	95.86	93.44	55.81
13	20.39	9.08	12.87	-63.15	-66.27	2.01	34.65	98.04	149.96	102.15	111.86	55.67	95.35
14	16.66	23.16	4.43	-34.94	-64.95	-12.66	13.06	86.88	195.43	107.88	91.42	125.19	73.16
15	13.15	-15.77	-31.36	-48.93	-80.31	-41.85	21.31	107.87	124.65	164.43	96.54	76.47	39.64
16	29.09	2.94	-9.24	-66.07	-32.47	-31.18	-14.70	119.76	229.46	197.22	89.34	121.06	104.55
17	44.47	11.75	-1.21	-37.77	-65.02	-4.98	5.92	101.73	213.98	124.35	127.05	103.67	149.56
18	76.16	18.94	24.00	-15.60	-67.52	-10.89	63.81	106.61	124.23	136.44	116.91	155.93	12.17
19	-19.60	53.45	-24.23	-38.99	-21.21	-7.70	53.20	160.01	113.77	170.11	108.14	113.18	6.52
20	-0.26	26.20	-25.25	-58.82	-51.20	-44.47	74.36	170.37	133.29	202.26	207.17	103.05	125.58
21	34.11	68.13	-59.70	-28.36	-3.94	-31.66	-31.87	139.22	154.97	141.36	112.84	93.43	40.20
22	35.68	24.17	-16.46	-24.98	-41.16	-38.76	-15.87	110.07	239.37	136.10	144.61	50.95	109.05
23	34.97	112.43	24.72	-11.55	-51.99	16.61	16.24	107.71	216.91	144.79	112.37	61.98	92.35
24	52.34	106.18	32.53	-15.61	-30.64	-18.99	35.28	71.28	175.76	128.00	175.35	143.52	78.57
25	43.01	-25.85	3.92	-37.98	-29.49	-21.72	84.37	94.70	161.48	160.89	127.37	127.37	39.36

Continued on next page.

Table 7-12 (continued). Lake Superior net basin supply (mm) transformed from 1948–94 historical meteorology with September 15, 1998, initial conditions.

Index, i (1)	Sep 98 (2)	Oct 98 (3)	Nov 98 (4)	Dec 98 (5)	Jan 99 (6)	Feb 99 (7)	Mar 99 (8)	Apr 99 (9)	May 99 (10)	Jun 99 (11)	Jul 99 (12)	Aug 99 (13)	Sep 99 (14)
26	37.62	30.38	-22.07	-33.48	-29.99	-24.82	-9.90	126.14	172.10	156.74	127.94	135.91	57.66
27	10.86	14.05	23.20	-37.68	-9.62	-20.05	-4.21	68.90	175.75	180.62	97.92	41.37	50.58
28	16.93	5.10	53.40	-37.16	-33.34	4.21	63.46	146.42	97.26	163.81	94.60	18.65	-14.21
29	-16.74	-39.25	-69.46	-67.62	-39.41	24.85	139.04	145.56	120.57	154.05	117.88	133.07	178.90
30	63.44	24.51	38.22	-20.18	-61.48	-33.41	-3.56	65.17	175.60	137.82	147.50	126.22	55.41
31	5.91	-12.56	-33.04	-62.24	-53.94	-5.25	89.42	117.42	224.00	182.09	121.48	100.33	62.57
32	2.44	76.06	2.00	-26.31	-2.09	-27.62	-7.73	98.54	100.54	107.11	116.39	106.25	104.69
33	54.34	0.06	-10.00	-41.01	-36.13	49.99	42.08	135.21	118.85	197.59	73.44	49.22	-6.84
34	-6.96	55.11	-22.71	-17.19	-53.47	-25.05	9.44	95.41	184.91	114.52	205.00	86.28	114.15
35	39.35	116.71	19.39	12.36	-21.27	-3.86	26.45	88.45	127.56	118.45	128.20	75.96	94.52
36	44.62	79.37	19.66	-45.88	-28.52	17.21	-3.18	100.75	155.38	199.09	130.74	100.71	65.23
37	31.67	51.80	-20.01	4.98	-47.68	-7.45	21.95	112.85	179.55	145.08	151.73	125.79	161.10
38	110.04	55.77	48.73	-51.87	-37.64	-10.04	32.87	154.38	121.38	175.12	159.12	102.57	97.37
39	50.09	28.62	-16.11	-38.97	-33.64	-17.05	13.38	50.94	118.25	68.43	127.47	71.17	54.35
40	26.45	3.80	-0.10	-0.06	-35.06	-32.49	27.76	90.71	133.02	119.09	101.48	180.64	62.28
41	57.52	20.97	85.98	-6.99	-1.59	-28.22	15.17	95.62	172.00	164.85	83.70	114.30	34.23
42	9.01	0.20	-20.75	-42.41	-5.85	-22.66	51.40	104.35	142.80	190.23	123.44	65.19	63.97
43	23.12	75.79	-13.79	-37.22	-21.50	-5.08	50.45	133.84	157.69	114.97	132.25	36.27	70.69
44	31.17	60.42	49.30	-19.26	-19.86	-5.47	-0.48	114.06	166.02	109.44	175.03	112.57	137.52
45	50.58	10.01	7.63	2.46	-30.75	-43.83	2.49	128.92	194.32	155.51	192.45	107.14	57.43
46	2.01	16.00	-14.26	-33.79	-49.56	-28.53	19.91	115.42	143.34	141.44	149.05	94.29	71.47
47	62.46	18.05	-0.71	-31.77	-43.18	-31.23	38.66	68.90	149.92	84.63	148.95	75.89	106.99

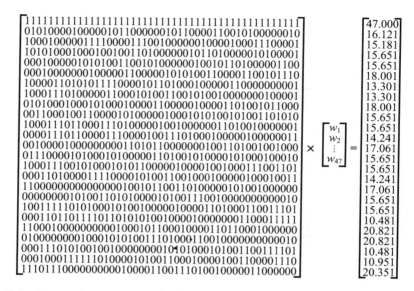

Figure 7-3. Alternative representation of Equation (6-3) and the first 24 equations in Figure 4-5.

differentials are set to zero, as in Equations (7-7) and (7-8). Table 7-15 presents the solution of Equation (7-16) found by using the *first* procedural algorithm in Figure 7-2 (using all scenarios). Table 7-16 presents the solution of Equation (7-16) found by using the *second* procedural algorithm in Figure 7-2 (maximizing use of meteorology outlooks). Checks reveal that $\sum w_i^2 < 2n$ for both procedural algorithms; therefore, the

Table 7-13. Outlook weights maximizing use of meteorology outlooks in Figure 7-3 for the Lake Superior supply outlook example.

Index, i (1)	Weight, w_i (2)	Index, i (3)	Weight, w_i (4)	Index, i (5)	Weight, w_i (6)
1	1.060475	17	1.766160	33	2.120593
2	2.768731	18	1.666037	34	2.734721
3	0	19	2.971981	35	1.360656
4	0	20	0.372594	36	0.453465
5	0	21	0.573141	37	1.114288
6	0.778769	22	3.028104	38	0.848485
7	0.743021	23	2.363683	39	0
8	0	24	2.020468	40	0.153503
9	0	25	0.086473	41	2.099952
10	0.007176	26	1.870353	42	1.595679
11	0	27	1.964278	43	0.351949
12	1.744236	28	0.966552	44	0
13	0	29	0.942909	45	0.424029
14	2.047606	30	0.275731	46	0.933164
15	0	31	0.087902	47	0
16	0	32	2.703132		

Table 7-14. September 15, 1998, Lake Superior probabilistic outlook of net basin supply (mm) maximizing use of meteorology outlooks.

Month	Nonexceedance quantiles								
	3%	10%	20%	30%	50%	70%	80%	90%	97%
(1)	(2)	(3)	(4)	(5)	(6)	(7)	(8)	(9)	(10)
Sep 98	−21.1	−13.9	−3.39	2.38	31.7	43.2	52.2	57.3	74.8
Oct 98	−33.6	−2.10	11.2	18.7	24.2	54.7	68.1	93.2	112
Nov 98	−61.5	−24.9	−22.7	−19.9	−11.3	19.0	24.1	33.8	64.1
Dec 98	−77.6	−53.0	−43.1	−39.6	−34.2	−23.0	−15.8	−10.6	4.87
Jan 99	−67.3	−65.0	−53.1	−49.0	−36.4	−29.6	−21.2	−3.87	−1.93
Feb 99	−40.9	−38.9	−31.1	−28.1	−20.8	−9.11	−5.00	12.5	33.3
Mar 99	−28.9	−18.5	−8.15	−4.34	11.7	27.2	41.5	52.9	66.1
Apr 99	68.9	71.3	83.8	95.3	105	122	135	155	161
May 99	97.8	105	120	133	172	195	212	233	243
Jun 99	103	108	117	129	143	165	175	197	201
Jul 99	71.6	83.0	95.6	107	118	139	148	175	201
Aug 99	41.0	46.6	53.4	66.3	93.4	113	123	135	147
Sep 99	−12.7	0.41	33.8	50.2	64.9	93.6	107	113	156

solutions in both cases represent minimums. (All computations are with probabilities, both reference quantiles and forecasts, significant to three digits after the decimal point.)

The first procedural algorithm matches Equations (4-9a) through (4-9g) while using all of the meteorology time series segments from 1948 to 1994; see Table 7-15. The second procedural algorithm matches Equations (4-9a) through (4-9h) but has zero weights for years 1952, 1953, 1957, 1963, 1972, 1974, 1982, and 1994; see Table 7-16. (It is interesting to note that none of the years omitted are La Niña years in Table 4-2.)

Table 7-15. Outlook weights using all meteorology time series segments for Equation (7-16) for the La Niña Lake Superior supply outlook example.

Index, i	Weight, w_i	Index, i	Weight, w_i	Index, i	Weight, w_i
(1)	(2)	(3)	(4)	(5)	(6)
1	0.259214	17	2.167816	33	0.565279
2	1.780569	18	0.259214	34	1.314670
3	0.945037	19	1.468592	35	0.259214
4	1.264503	20	0.644569	36	0.945037
5	0.259214	21	0.964035	37	0.091891
6	0.259214	22	1.081345	38	0.945037
7	0.259214	23	1.780569	39	0.964035
8	0.565279	24	2.013894	40	0.259214
9	1.729262	25	0.245813	41	0.093670
10	0.259214	26	2.718715	42	0.945037
11	2.167816	27	0.259214	43	1.116178
12	0.413135	28	0.796713	44	0.413135
13	0.796713	29	1.883183	45	0.964035
14	1.030038	30	2.019491	46	2.167816
15	2.013894	31	2.167816	47	0.259214
16	0.259214	32	0.964035		

Table 7-16. Outlook weights maximizing use of meteorology event probabilities for Equation (7-16) for the La Niña Lake Superior supply outlook example.

Index, i (1)	Weight, w_i (2)	Index, i (3)	Weight, w_i (4)	Index, i (5)	Weight, w_i (6)
1	0.599708	17	2.209499	33	0.595273
2	1.798767	18	0.599708	34	1.244683
3	1.287485	19	1.495376	35	0
4	1.309397	20	0.651246	36	1.287485
5	0	21	1.340538	37	0.322668
6	0	22	1.084643	38	0.620105
7	0.599708	23	1.798767	39	0.673159
8	0.595273	24	1.958807	40	0.599708
9	1.937663	25	0	41	0.161108
10	0	26	2.699638	42	0.620105
11	2.209499	27	0	43	1.085412
12	0.850400	28	1.063499	44	0.183021
13	0.396120	29	1.520976	45	0.673159
14	1.223539	30	1.985514	46	2.209499
15	1.958807	31	2.209499	47	0
16	0	32	1.340538		

Using either set of weights allows probabilistic hydrology outlooks for Lake Superior net basin supply to be built from Table 7-12, in the way they were built for Table 7-14 in the previous example. See Exercise A2-7 in Appendix 2 for this example, which also illustrates the setting of arbitrary user-defined (non-agency) probabilities such as Equations (4-9).

ORDERING PRIORITIES
There are several practical ways for ordering priorities. First, a practitioner would use meteorology probability forecasts of appropriate lead and length for the derivative forecasts at hand. Thus, one would place meteorology forecasts over the next few days at higher priority than a 1-month forecast if one desired the derivative hydrology forecast at the end of the week. Likewise, if a lake level outlook over the next 6 months was to be made, then the 3-month meteorology forecasts beginning with the present month, the following month, and the month after would be more important than the second-week meteorology forecast. Another consideration for the practitioner is to place the most important variables first, reflecting his or her goals or purposes. For example, February air temperatures may be much more important for snowmelt events than June-July-August precipitation. Users may also assign priorities according to their confidence in the meteorology outlooks. For example, an older meteorology forecast may have a much lower priority to the user than a more recent one. Or one agency may have a better forecast success rate in the user's application area than another agency; this too can be reflected in the user's priority listing.

Very often, priorities do not change much in day-to-day forecasting activities. A lot of thought may go into selecting a reasonable set of priorities for the agency forecasts that are used to make a derivative hydrology outlook. As long as the meteorological outlooks that are being used are not removed (even though they are allowed to change) from day to day, their priorities may remain unchanged. It is necessary only to recalculate the weights when the meteorology forecasts or their priorities change. The same set of weights can be used in day-to-day updated hydrology forecasts in the interim, reflecting only updated initial conditions used in the model simulations.

ADDITIONAL METHODOLOGY CONSIDERATIONS

Formulating an optimization, as described in this chapter, allows for a general approach in determining weights in the face of multiple outlooks. However, this formulation also involves arbitrary choices, the largest of which is the selection of a relevant objective function. As mentioned earlier, other measures of relevance of the weights to a goal are possible and could require reformulating the solution methodology. An earlier approach, not described in this book, was to minimize the sum of squared differences between the relative frequencies associated with the bivariate distribution of precipitation and temperature before and after application of the weights. The goal was to make the resulting joint distribution as similar as possible to that observed historically while making the marginal distributions match the climate outlooks. Unfortunately, that method was intractable for consideration of more than one climate outlook. Alternative formulations that use linear measures for comparing alternative solutions to determine which is "best" are described in Chapter 10.

Most significantly, the method allows joint consideration of multiple probabilistic meteorology outlooks of event probabilities. The next chapter extends the methodology to also include probabilistic meteorology outlooks of most-probable events.

Chapter 8

MIXING MOST-PROBABLE METEOROLOGY OUTLOOKS

Chapters 6 and 7 discussed restructuring of the operational hydrology future scenarios sample to match forecast event probabilities as given, for example, in the National Oceanic and Atmospheric Administration's (NOAA's) monthly *Climate Outlooks* or its 8–14 day outlook. However, that chapter did not address matching most-probable event forecasts such as the NOAA 6–10 day outlook or the Environment Canada (EC) 1-month and 3-month outlooks. This chapter extends the approach to mix all probabilistic meteorology outlooks to generate hydrology outlooks.

MATCHING MOST-PROBABLE EVENTS

Consider matching most-probable event forecasts such as are available in NOAA's 6–10 day outlooks for average air temperature and total precipitation, EC's monthly outlooks for average air temperature, or EC's seasonal and extended seasonal outlooks for average air temperature and total precipitation. Most-probable event forecasts are a special case of a more general category of probability statements. Generally, $r + 1$ intervals for a variable's values are set by defining interval limits, $z_1 < z_2 \cdots < z_r$. The general form of the probability statement in which a most-probable event forecast can be cast is that the jth event (interval) has a probability in excess of a specified value; the probability can be written in terms of the relative frequencies to be matched:

$$\hat{P}\left[z_{j-1} < X \leq z_j\right] > \phi_j \qquad (8\text{-}1)$$

where X may be average air temperature or total precipitation and ϕ_j is a probability limit. $z_0 = -\infty$ and $z_{r+1} = +\infty$ are understood and for these cases, Equation (8-1) becomes

$$\hat{P}\left[z_0 < X \leq z_1\right] = \hat{P}\left[X \leq z_1\right] > \phi_1 \qquad (8\text{-}2a)$$

$$\hat{P}\left[z_r < X \leq z_{r+1}\right] = \hat{P}\left[X > z_r\right] > \phi_{r+1} \qquad (8\text{-}2b)$$

[In the NOAA forecast of most-probable air temperature and precipitation events and in the EC forecasts of most-probable precipitation events, z_k is defined as the γ_k quantile (ξ_k) estimated from the 1961–90 period. In the EC forecasts of most-probable air temperature events, the quantiles are estimated from the 1963–93 period. In general:

$$\hat{P}\left[X \leq \hat{\xi}_k\right] = \gamma_k \qquad 1 \leq k \leq r \qquad (8\text{-}3)$$

where $\gamma_1 < \gamma_2 < \cdots < \gamma_r$ and ϕ_k is defined in terms of the quantile probabilities:

$$\phi_k = \gamma_k - \gamma_{k-1} \qquad 1 \leq k \leq r+1 \qquad (8\text{-}4)$$

where $\gamma_0 = 0$ and $\gamma_{r+1} = 1$. For the NOAA 6–10 day most-probable event temperature forecast, $r = 4$, $\gamma_1 = 0.1$, $\gamma_2 = 0.3$, $\gamma_3 = 0.7$, and $\gamma_4 = 0.9$ ($\phi_1 = 0.1$, $\phi_2 = 0.2$, $\phi_3 = 0.4$, $\phi_4 = 0.2$, and $\phi_5 = 0.1$); for the NOAA 6–10 day most-probable event precipitation forecast and all of the EC most-probable event temperature and precipitation forecasts, $r = 2$, $\gamma_1 = 1/3$, and $\gamma_2 = 2/3$ ($\phi_1 = \phi_2 = \phi_3 = 1/3$). However, the more general definitions of z_k and ϕ_k are used in this chapter to allow for other outlooks that may be more broadly defined than either of the present NOAA or EC most-probable event forecasts.]

Many most-probable event forecasts are accompanied by the implicit assumption that *only* the most-probable interval has forecast probability exceeding its reference probability. Equation (8-1) would then become:

$$\hat{P}\left[z_{j-1} < X \leq z_j\right] > \phi_j \tag{8-5a}$$

$$\hat{P}\left[z_{k-1} < X \leq z_k\right] \leq \phi_k \qquad k = 1, \ldots, r+1; \quad k \neq j \tag{8-5b}$$

Alternatively, Equations (8-5) can be written in terms of the complement for the first event as:

$$\hat{P}\left[\text{not}\left(z_{j-1} < X \leq z_j\right)\right] < 1 - \phi_j \tag{8-6a}$$

$$\hat{P}\left[z_{k-1} < X \leq z_k\right] \leq \phi_k \qquad k = 1, \ldots, r+1; \quad k \neq j \tag{8-6b}$$

If the user does not wish to make the assumption, then the r inequalities in Equations (8-6b) can be omitted.

Weights are determined by matching relative frequencies in the operational hydrology sample to the most-probable interval forecast of Equations (8-6) [as was done in Chapter 6 to replace Equation (6-1) with Equation (6-2) by using Equation (5-8)]:

$$\frac{1}{n} \sum_{i \mid \text{not}\left(z_{j-1} < x_i \leq z_j\right)} w_i < 1 - \phi_j \tag{8-7a}$$

$$\frac{1}{n} \sum_{i \mid z_{k-1} < x_i \leq z_k} w_i \leq \phi_k \qquad k = 1, \ldots, r+1; \quad k \neq j \tag{8-7b}$$

Alternatively, write Equations (8-7) as follows:

$$\sum_{i=1}^{n} \alpha_{j,i} w_i < e_j \tag{8-8a}$$

$$\sum_{i=1}^{n} \alpha_{k,i} w_i \leq e_k \qquad k = 1, \ldots, r+1; \quad k \neq j \tag{8-8b}$$

where the $\alpha_{k,i}$ are defined, as they were for Equation (6-10), as 1 or 0, corresponding to the inclusion or exclusion, respectively, of each variable in the respective appropriate sets of Equations (8-7), and where e_k corresponds to the probability limits specified in the most-probable event forecast [$e_j = n\left(1 - \phi_j\right)$ and $e_k = n\phi_k$, $k \neq j$]. The $r + 1$ inequalities in Equations (8-8) represent one most-probable event forecast; if there are multiple most-probable event forecasts (from different agencies, for different periods and lags, and for different variables), represent them by the $p + q$ inequalities:

$$\sum_{i=1}^{n} \alpha_{k,i} w_i \; < \; e_k \qquad\qquad k = 1, \ldots, p \qquad\qquad\qquad (8\text{-}9a)$$

$$\sum_{i=1}^{n} \alpha_{k,i} w_i \; \leq \; e_k \qquad\qquad k = p+1, \ldots, p+q \qquad\quad (8\text{-}9b)$$

where p = the total number of strictly-less-than constraints and q = the total number of less-than-or-equal-to constraints to be considered. Note that while Equations (8-9) may refer to different variables over different periods with different lengths and lag times, the equations are written in terms of a single set of weights (w_i, $i = 1, \ldots, n$) as was done for Equation (6-10).

MIXING PROBABILISTIC METEOROLOGY OUTLOOKS
By adding the constraints corresponding to most-probable event forecasts in Equations (8-9) to those of the event probability forecasts in Equation (6-10) and the requirement of Equation (6-3), the following set of equations is formed to be to solved simultaneously:

$$\sum_{i=1}^{n} \alpha_{k,i} w_i \; = \; e_k \qquad\qquad k = 1, \ldots, m \qquad\qquad\qquad (8\text{-}10a)$$

$$\sum_{i=1}^{n} \alpha_{k,i} w_i \; < \; e_k \qquad\qquad k = m+1, \ldots, m+p \qquad\quad (8\text{-}10b)$$

$$\sum_{i=1}^{n} \alpha_{k,i} w_i \; \leq \; e_k \qquad\qquad k = m+p+1, \ldots, m+p+q \quad (8\text{-}10c)$$

Again defining an optimization problem and solving by searching for an "optimum" solution, as in Equations (7-5), the optimization becomes:

$$\min \sum_{i=1}^{n} (w_i - 1)^2 \text{ subject to} \qquad\qquad\qquad\qquad\qquad (8\text{-}11a)$$

$$\sum_{i=1}^{n} \alpha_{k,i} w_i \; = \; e_k \qquad\qquad k = 1, \ldots, m \qquad\qquad\qquad (8\text{-}11b)$$

$$\sum_{i=1}^{n} \alpha_{k,i} w_i \; < \; e_k \qquad\qquad k = m+1, \ldots, m+p \qquad\quad (8\text{-}11c)$$

$$\sum_{i=1}^{n} \alpha_{k,i} w_i \; \leq \; e_k \qquad\qquad k = m+p+1, \ldots, m+p+q \quad (8\text{-}11d)$$

The solution to Equations (8-11) may give positive, zero, or negative weights, but only nonnegative weights make physical sense. Again, two procedural algorithms are used for finding nonnegative weights without adding additional constraints to Equations (8-11), so that the solution is analytically tractable. These algorithms repeatedly eliminate the lowest-priority equation or inequality in Equations (8-11b), (8-11c), and (8-11d) until nonnegative weights are obtained. As before, the first algorithm guarantees that all scenarios in the operational hydrology sample are used and the second maximizes the number of equations or inequalities (meteorology outlooks) used.

Equations (8-11) are equivalent to:

$$\min \sum_{i=1}^{n}(w_i - 1)^2 \text{ subject to} \tag{8-12a}$$

$$\sum_{i=1}^{n}\alpha_{k,i}w_i = e_k \qquad k = 1, \ldots, m \tag{8-12b}$$

$$\sum_{i=1}^{n}\alpha_{k,i}w_i + w_{n+k-m} = e_k \qquad k = m+1, \ldots, m+p+q \tag{8-12c}$$

$$w_i > 0 \qquad i = n+1, \ldots, n+p \tag{8-12d}$$

$$w_i \geq 0 \qquad i = n+p+1, \ldots, n+p+q \tag{8-12e}$$

where the w_i, $(i = n + 1, \ldots, n + p + q)$ are "slack" variables added to change considera-tion of an inequality constraint to consideration of an equality constraint in the optimi-zation. This, in turn, is equivalent to:

$$\min \sum_{i=1}^{n}(w_i - 1)^2 \qquad \text{subject to} \tag{8-13a}$$

$$\sum_{i=1}^{n+p+q}\alpha_{k,i}w_i = e_k \qquad k = 1, \ldots, m+p+q \tag{8-13b}$$

$$w_i > 0 \qquad i = n+1, \ldots, n+p \tag{8-13c}$$

$$w_i \geq 0 \qquad i = n+p+1, \ldots, n+p+q \tag{8-13d}$$

where the additional coefficients are defined as follows:

$$\alpha_{k,i} = 0 \qquad k = 1, \ldots, m \qquad\qquad i = n+1, \ldots, n+p+q \tag{8-14a}$$

$$\alpha_{k,i} = 1 \qquad k = m+1, \ldots, m+p+q \qquad i = n+k-m \tag{8-14b}$$

$$\alpha_{k,i} = 0 \qquad k = m+1, \ldots, m+p+q \qquad i > n, \ i \neq n+k-m \tag{8-14c}$$

If the non-negativity constraints $(w_i > 0$, $i = n + 1, \ldots, n+p$ and $w_i \geq 0$, $i = n + p + 1, \ldots, n + p + q)$ are ignored for now, Equations (8-13) become:

$$\min \sum_{i=1}^{n}(w_i - 1)^2 \text{ subject to} \tag{8-15a}$$

$$\sum_{i=1}^{n+p+q}\alpha_{k,i}w_i = e_k \qquad k = 1, \ldots, m+p+q \tag{8-15b}$$

which is similar to Equations (7-5) and may be solved as before (Croley 1996, 1997a) by defining the Lagrangian (Hillier and Lieberman 1969, pp. 603-08),

$$L = \sum_{i=1}^{n}(w_i - 1)^2 - \sum_{k=1}^{m+p+q}\lambda_k\left(\sum_{i=1}^{n+p+q}\alpha_{k,i}w_i - e_k\right) \tag{8-16}$$

where λ_k = the unit penalty of violating the kth constraint in the optimization, and by setting the first derivatives with respect to each variable to zero:

$$\frac{\partial L}{\partial w_i} = 2(w_i - 1) - \sum_{k=1}^{m+p+q} \lambda_k \alpha_{k,i} = 0 \qquad i = 1, \ldots, n \tag{8-17a}$$

$$\frac{\partial L}{\partial w_i} = -\sum_{k=1}^{m+p+q} \lambda_k \alpha_{k,i} \qquad\qquad = 0 \qquad i = n+1, \ldots, n+p+q \tag{8-17b}$$

$$\frac{\partial L}{\partial \lambda_k} = -\sum_{i=1}^{n+p+q} \alpha_{k,i} w_i + e_k \qquad = 0 \qquad k = 1, \ldots, m+p+q \tag{8-17c}$$

This is a set of necessary but not sufficient conditions for the minimization of Equation (8-16) or the problem of Equations (8-15). The solution represents a "critical" point and must be checked further to identify it as either a minimum or a maximum. Equations (8-17) are linear and solvable via the Gauss-Jordan method of elimination because there are $m + n + 2p + 2q$ equations in $m + n + 2p + 2q$ unknowns (same number of equations and variables). For this problem where one of the equations in Equations (6-10) and (8-11) is Equation (6-3), the solution of Equations (8-17) represents the minimum if $\sum w_i^2 < 2n$ and the maximum if $\sum w_i^2 > 2n$ (see Appendix 3). Note that these are the same sufficiency conditions as for Equations (7-7). Equations (8-17) can be written in vector form as in Figure 8-1.

The solution of Equations (8-15) may give positive, zero, or negative weights and slack variables, but only nonnegative or strictly positive weights (either $w_i \geq 0$ or $w_i > 0$, $i = 1, \ldots, n$) and slack variables ($w_i > 0$, $i = n+1, \ldots, n+p$ and $w_i \geq 0$, $i = n+p + 1, \ldots, n+p+q$) make physical sense, and the optimization must be further constrained. Two cases arise here:

$$w_i > 0 \qquad i = 1, \ldots, n \tag{8-18a}$$
$$w_i > 0 \qquad i = n+1, \ldots, n+p \tag{8-18b}$$
$$w_i \geq 0 \qquad i = n+p+1, \ldots, n+p+q \tag{8-18c}$$

and

$$w_i \geq 0 \qquad i = 1, \ldots, n \tag{8-19a}$$
$$w_i > 0 \qquad i = n+1, \ldots, n+p \tag{8-19b}$$
$$w_i \geq 0 \qquad i = n+p+1, \ldots, n+p+q \tag{8-19c}$$

In both cases, there is a mixture of strictly positive ($w_i > 0$) and simply nonnegative ($w_i \geq 0$) weights and slack variables for the optimization. These additional constraints can result in infeasibility (meaning there is no solution), and equations must be eliminated from Equations (8-15) to allow a feasible solution. To facilitate this, the engineer or hydrologist must prioritize the probabilistic meteorology outlook equations [and, hence, the equations in Equations (8-15)] so that the least important ones (lowest priority) can be eliminated first. The equation in Equations (8-15b) corresponding to Equation (6-3) should always be given top priority.

A procedural algorithm of successive optimizations is depicted in Figure 8-2; it preserves as many of the probability equations as possible while yielding results identical to Figure 7-2 when no slack variables are present (Croley 1997b). In Figure 8-2, if simple nonnegativity conditions would be violated in an optimization, even though other positivity conditions may also be violated, the procedural algorithm adds a zero constraint ($w_i = 0$) for each negative variable ($w_i < 0$), as long as the resulting equation set still represents a nonempty space, and it solves the optimization again. If the resulting equation set would represent an empty solution space, then the algorithm eliminates all ear-

$$
\begin{bmatrix}
-2 & 0 & \cdots & 0 & 0 & 0 & \cdots & 0 & \alpha_{1,1} & \alpha_{2,1} & \cdots & \alpha_{m+p+q,1} \\
0 & -2 & \cdots & 0 & 0 & 0 & \cdots & 0 & \alpha_{1,2} & \alpha_{2,2} & \cdots & \alpha_{m+p+q,2} \\
\vdots & \vdots & & \vdots & \vdots & \vdots & & \vdots & \vdots & \vdots & & \vdots \\
0 & 0 & \cdots & -2 & 0 & 0 & \cdots & 0 & \alpha_{1,n} & \alpha_{2,n} & \cdots & \alpha_{m+p+q,n} \\
0 & 0 & \cdots & 0 & 0 & 0 & \cdots & 0 & \alpha_{1,n+1} & \alpha_{2,n+1} & \cdots & \alpha_{m+p+q,n+1} \\
0 & 0 & \cdots & 0 & 0 & 0 & \cdots & 0 & \alpha_{1,n+2} & \alpha_{2,n+2} & \cdots & \alpha_{m+p+q,n+2} \\
\vdots & \vdots & & \vdots & \vdots & \vdots & & \vdots & \vdots & \vdots & & \vdots \\
0 & 0 & \cdots & 0 & 0 & 0 & \cdots & 0 & \alpha_{1,n+p+q} & \alpha_{2,n+p+q} & \cdots & \alpha_{m+p+q,n+p+q} \\
\alpha_{1,1} & \alpha_{1,2} & \cdots & \alpha_{1,n} & \alpha_{1,n+1} & \alpha_{1,n+2} & \cdots & \alpha_{1,n+p+q} & 0 & 0 & \cdots & 0 \\
\alpha_{2,1} & \alpha_{2,2} & \cdots & \alpha_{2,n} & \alpha_{2,n+1} & \alpha_{2,n+2} & \cdots & \alpha_{2,n+p+q} & 0 & 0 & \cdots & 0 \\
\vdots & \vdots & & \vdots & \vdots & \vdots & & \vdots & \vdots & \vdots & & \vdots \\
\alpha_{m+p+q,1} & \alpha_{m+p+q,2} & \cdots & \alpha_{m+p+q,n} & \alpha_{m+p+q,n+1} & \alpha_{m+p+q,n+2} & \cdots & \alpha_{m+p+q,n+p+q} & 0 & 0 & \cdots & 0
\end{bmatrix}
\begin{bmatrix}
w_1 \\ w_2 \\ \vdots \\ w_n \\ w_{n+1} \\ w_{n+2} \\ \vdots \\ w_{n+p+q} \\ \lambda_1 \\ \lambda_2 \\ \vdots \\ \lambda_{m+p+q}
\end{bmatrix}
=
\begin{bmatrix}
-2 \\ -2 \\ \vdots \\ -2 \\ 0 \\ 0 \\ \vdots \\ 0 \\ e_1 \\ e_2 \\ \vdots \\ e_{m+p+q}
\end{bmatrix}
$$

Figure 8-1. Necessary conditions for optimization matching mixed probabilistic meteorology outlooks.

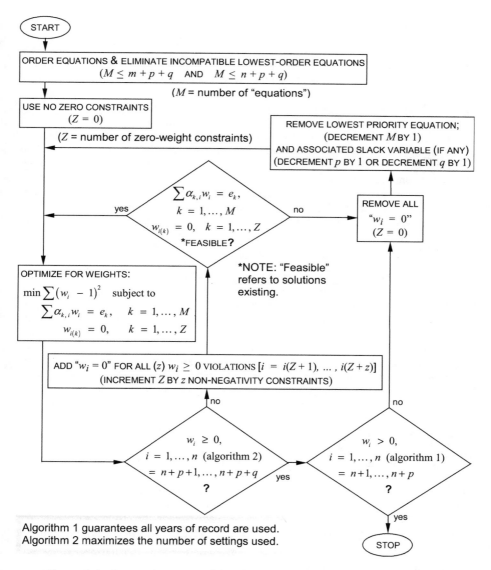

Figure 8-2. Determining physically relevant weights and slack variables.

lier-added zero constraints and the lowest-priority probability equation instead and solves the optimization again. If only positivity constraints would be violated, then the algorithm simply eliminates all earlier-added zero constraints and the lowest-priority probability equation and solves the optimization again. Two variations are depicted in Figure 8-2. Algorithm 1 guarantees that only strictly positive weights will result, as in Equations (8-18), and all possible future scenarios will be used (no scenario is weighted by zero and effectively eliminated). Algorithm 2 disallows some of the possible future scenarios (by allowing zero weights), as in Equations (8-19); this generally allows satisfaction of more probability equations than the first algorithm.

MIXED MULTIPLE OUTLOOKS EXAMPLE

In generating a Lake Superior basin hydrology outlook on September 15, 1998, probabilistic outlooks made on five separate dates are available:

1. The NOAA *Climate Outlook* for September 1998 was made August 13, 1998, and consists of event probabilities for September air temperature and precipitation and for 3-month air temperature and precipitation over 13 periods, successively lagged 1 month each, starting with September-October-November 1998. The first six outlook periods are to be used (September, SON, OND, NDJ, DJF, and JFM). The outlook was summarized for the Lake Superior basin area in Figure 4-5 and calls for lower than normal air temperatures in September and SON, followed by normal air temperatures until the JFM period, when they again are expected to be low. While it also calls for normal September precipitation, higher than normal precipitation is expected for all 3-month periods from SON through JFM.

2. The NOAA 8–14 day outlook for September 11–17, 1998 (event probabilities for second-week air temperature and precipitation), was made September 3, 1998. It is summarized in Equations (4-4) for the Lake Superior basin and calls for higher temperatures and lower precipitation than normal for the week of September 11–17.

3. The NOAA 6–10 day outlook for September 17–21, 1998 (most-probable event for 5-day air temperature and precipitation), was made September 11, 1998. It is summarized in Equations (4-10) for the Lake Superior basin and calls for higher than normal temperatures and normal precipitation during September 17–21.

4. The EC climate outlook for September 1998 (most-probable event for September 1998 air temperature) was made August 31, 1998. Summarized in Equations (4-12) for the Lake Superior basin, it calls for lower than normal September temperatures.

5. The EC climate outlook for September-October-November 1998 (most-probable event for 3-month air temperature and precipitation) was made August 31, 1998. It is summarized in Equations (4-13) for the Lake Superior basin and calls for normal SON air temperatures and higher than normal SON precipitation.

All of these outlooks are summarized in Figure 8-3.

While these outlooks are largely compatible with each other, there are a few obviously impossible joint forecasts that make for an infeasible solution. The fifth event probability equation from the NOAA 1- and 3-month outlook in Figure 8-3 and the first most-probable event equation from the EC seasonal outlook directly conflict; both cannot be satisfied (i.e., the probability that $T_{SON98} \le \hat{\tau}_{SON, 0.333}$ *cannot* both be equal to 0.383 *and* be less than or equal to 0.333). Likewise, the fifth and sixth equations from the NOAA 1- and 3-month outlooks and the second equation from the EC seasonal outlook also directly conflict. (That is, the fifth and sixth equations from the NOAA 1- and 3-month outlook together imply that $\hat{P}\left[\hat{\tau}_{SON, 0.333} < T_{SON98} \le \hat{\tau}_{SON, 0.667}\right] = 0.334$, because these three equations must sum to unity. The second equation from the EC seasonal outlook says this probability must be greater than 0.334.) Therefore, the EC seasonal outlook for air temperature is removed from further consideration in this example and is shaded in Figure 8-3. The remaining 42 equations could be used. However, even though the second equation of the EC monthly outlook is not in conflict with the first and second equations of the NOAA 1- and 3-month outlook in Figure 8-3, its removal allows more equations to be satisfied; it is also shaded in Figure 8-3. This sometimes

NOAA 1- & 3-Month Climate Outlook Event Probabilities

$\hat{P}\left[T_{\text{Sep98}} \le \hat{\tau}_{\text{Sep},0.333}\right] = 0.343$

$\hat{P}\left[T_{\text{Sep98}} > \hat{\tau}_{\text{Sep},0.667}\right] = 0.323$

$\hat{P}\left[Q_{\text{Sep98}} \le \hat{\theta}_{\text{Sep},0.333}\right] = 0.333$

$\hat{P}\left[Q_{\text{Sep98}} > \hat{\theta}_{\text{Sep},0.667}\right] = 0.333$

$\hat{P}\left[T_{\text{SON98}} \le \hat{\tau}_{\text{SON},0.333}\right] = 0.383$

$\hat{P}\left[T_{\text{SON98}} > \hat{\tau}_{\text{SON},0.667}\right] = 0.283$

$\hat{P}\left[Q_{\text{SON98}} \le \hat{\theta}_{\text{SON},0.333}\right] = 0.283$

$\hat{P}\left[Q_{\text{SON98}} > \hat{\theta}_{\text{SON},0.667}\right] = 0.383$

$\hat{P}\left[T_{\text{OND98}} \le \hat{\tau}_{\text{OND},0.333}\right] = 0.333$

$\hat{P}\left[T_{\text{OND98}} > \hat{\tau}_{\text{OND},0.667}\right] = 0.333$

$\hat{P}\left[Q_{\text{OND98}} \le \hat{\theta}_{\text{OND},0.333}\right] = 0.303$

$\hat{P}\left[Q_{\text{OND98}} > \hat{\theta}_{\text{OND},0.667}\right] = 0.363$

$\hat{P}\left[T_{\text{NDJ98}} \le \hat{\tau}_{\text{NDJ},0.333}\right] = 0.333$

$\hat{P}\left[T_{\text{NDJ98}} > \hat{\tau}_{\text{NDJ},0.667}\right] = 0.333$

$\hat{P}\left[Q_{\text{NDJ98}} \le \hat{\theta}_{\text{NDJ},0.333}\right] = 0.303$

$\hat{P}\left[Q_{\text{NDJ98}} > \hat{\theta}_{\text{NDJ},0.667}\right] = 0.363$

$\hat{P}\left[T_{\text{DJF98}} \le \hat{\tau}_{\text{DJF},0.333}\right] = 0.333$

$\hat{P}\left[T_{\text{DJF98}} > \hat{\tau}_{\text{DJF},0.667}\right] = 0.333$

$\hat{P}\left[Q_{\text{DJF98}} \le \hat{\theta}_{\text{DJF},0.333}\right] = 0.223$

$\hat{P}\left[Q_{\text{DJF98}} > \hat{\theta}_{\text{DJF},0.667}\right] = 0.443$

$\hat{P}\left[T_{\text{JFM99}} \le \hat{\tau}_{\text{JFM},0.333}\right] = 0.443$

$\hat{P}\left[T_{\text{JFM99}} > \hat{\tau}_{\text{JFM},0.667}\right] = 0.223$

$\hat{P}\left[Q_{\text{JFM99}} \le \hat{\theta}_{\text{JFM},0.333}\right] = 0.233$

$\hat{P}\left[Q_{\text{JFM99}} > \hat{\theta}_{\text{JFM},0.667}\right] = 0.433$

NOAA 8–14 Day Outlook Event Probabilities

$\hat{P}\left[T_{\text{11-17Sep98}} \le \hat{\tau}_{\text{11-17Sep},0.333}\right] = 0.153$

$\hat{P}\left[T_{\text{11-17Sep98}} > \hat{\tau}_{\text{11-17Sep},0.667}\right] = 0.513$

$\hat{P}\left[Q_{\text{11-17Sep98}} \le \hat{\theta}_{\text{11-17Sep},0.333}\right] = 0.393$

$\hat{P}\left[Q_{\text{11-17Sep98}} > \hat{\theta}_{\text{11-17Sep},0.667}\right] = 0.273$

NOAA 6–10 Day Outlook Most-Probable Events

$\hat{P}\left[T_{\text{17-21Sep98}} \le \hat{\tau}_{\text{17-21Sep},0.100}\right] \le 0.100$

$\hat{P}\left[\hat{\tau}_{\text{17-21Sep},0.100} < T_{\text{17-21Sep98}} \le \hat{\tau}_{\text{17-21Sep},0.300}\right] \le 0.200$

$\hat{P}\left[\hat{\tau}_{\text{17-21Sep},0.300} < T_{\text{17-21Sep98}} \le \hat{\tau}_{\text{17-21Sep},0.700}\right] \le 0.400$

$\hat{P}\left[\hat{\tau}_{\text{17-21Sep},0.700} < T_{\text{17-21Sep98}} \le \hat{\tau}_{\text{17-21Sep},0.900}\right] > 0.200$

$\hat{P}\left[T_{\text{17-21Sep98}} > \hat{\tau}_{\text{17-21Sep},0.900}\right] \le 0.100$

$\hat{P}\left[Q_{\text{17-21Sep98}} \le \hat{\theta}_{\text{17-21Sep},0.333}\right] \le 0.333$

$\hat{P}\left[\begin{array}{c}\hat{\theta}_{\text{17-21Sep},0.333} \\ < Q_{\text{17-21Sep98}} \le \\ \hat{\theta}_{\text{17-21Sep},0.667}\end{array}\right] > 0.334$

$\hat{P}\left[Q_{\text{17-21Sep98}} > \hat{\theta}_{\text{17-21Sep},0.667}\right] \le 0.333$

Environment Canada Monthly Most Probable Event

$\hat{P}\left[T_{\text{Sep98}} \le \hat{\tau}_{\text{Sep},0.333}\right] > 0.333$

$\hat{P}\left[\hat{\tau}_{\text{Sep},0.333} < T_{\text{Sep98}} \le \hat{\tau}_{\text{Sep},0.667}\right] \le 0.334$

$\hat{P}\left[T_{\text{Sep98}} > \hat{\tau}_{\text{Sep},0.667}\right] \le 0.333$

Environment Canada Seasonal Most-Probable Events

$\hat{P}\left[T_{\text{SON98}} \le \hat{\tau}_{\text{SON},0.333}\right] \le 0.333$

$\hat{P}\left[\hat{\tau}_{\text{SON},0.333} < T_{\text{SON98}} \le \hat{\tau}_{\text{SON},0.667}\right] > 0.334$

$\hat{P}\left[T_{\text{SON98}} > \hat{\tau}_{\text{SON},0.667}\right] \le 0.333$

$\hat{P}\left[Q_{\text{SON98}} \le \hat{\theta}_{\text{SON},0.333}\right] \le 0.333$

$\hat{P}\left[\hat{\theta}_{\text{SON},0.333} < Q_{\text{SON98}} \le \hat{\theta}_{\text{SON},0.667}\right] \le 0.334$

$\hat{P}\left[Q_{\text{SON98}} > \hat{\theta}_{\text{SON},0.667}\right] > 0.333$

Figure 8-3. Lake Superior probabilistic meteorology outlooks available September 15, 1998.

1	$\hat{P}\left[T_{11-17\text{Sep}98} \leq \hat{\tau}_{11-17\text{Sep},0.333}\right] = 0.153$	25 $\hat{P}\left[Q_{\text{SON}98} > \hat{\theta}_{\text{SON},0.667}\right] > 0.333$
2	$\hat{P}\left[T_{11-17\text{Sep}98} > \hat{\tau}_{11-17\text{Sep},0.667}\right] = 0.513$	26 $\hat{P}\left[T_{\text{OND}98} \leq \hat{\tau}_{\text{OND},0.333}\right] = 0.333$
3	$\hat{P}\left[Q_{11-17\text{Sep}98} \leq \hat{\theta}_{11-17\text{Sep},0.333}\right] = 0.393$	27 $\hat{P}\left[T_{\text{OND}98} > \hat{\tau}_{\text{OND},0.667}\right] = 0.333$
4	$\hat{P}\left[Q_{11-17\text{Sep}98} > \hat{\theta}_{11-17\text{Sep},0.667}\right] = 0.273$	28 $\hat{P}\left[Q_{\text{OND}98} \leq \hat{\theta}_{\text{OND},0.333}\right] = 0.303$
5	$\hat{P}\left[T_{17-21\text{Sep}98} \leq \hat{\tau}_{17-21\text{Sep},0.100}\right] \leq 0.100$	29 $\hat{P}\left[Q_{\text{OND}98} > \hat{\theta}_{\text{OND},0.667}\right] = 0.363$
6	$\hat{P}\left[\hat{\tau}_{17-21\text{Sep},0.100} < T_{17-21\text{Sep}98} \leq \hat{\tau}_{17-21\text{Sep},0.300}\right] \leq 0.200$	30 $\hat{P}\left[T_{\text{NDJ}98} \leq \hat{\tau}_{\text{NDJ},0.333}\right] = 0.333$
7	$\hat{P}\left[\hat{\tau}_{17-21\text{Sep},0.300} < T_{17-21\text{Sep}98} \leq \hat{\tau}_{17-21\text{Sep},0.700}\right] \leq 0.400$	31 $\hat{P}\left[T_{\text{NDJ}98} > \hat{\tau}_{\text{NDJ},0.667}\right] = 0.333$
8	$\hat{P}\left[\hat{\tau}_{17-21\text{Sep},0.700} < T_{17-21\text{Sep}98} \leq \hat{\tau}_{17-21\text{Sep},0.900}\right] > 0.200$	32 $\hat{P}\left[Q_{\text{NDJ}98} \leq \hat{\theta}_{\text{NDJ},0.333}\right] = 0.303$
9	$\hat{P}\left[T_{17-21\text{Sep}98} > \hat{\tau}_{17-21\text{Sep},0.900}\right] \leq 0.100$	33 $\hat{P}\left[Q_{\text{NDJ}98} > \hat{\theta}_{\text{NDJ},0.667}\right] = 0.363$
10	$\hat{P}\left[Q_{17-21\text{Sep}98} \leq \hat{\theta}_{17-21\text{Sep},0.333}\right] \leq 0.333$	34 $\hat{P}\left[T_{\text{DJF}98} \leq \hat{\tau}_{\text{DJF},0.333}\right] = 0.333$
11	$\hat{P}\left[\hat{\theta}_{17-21\text{Sep},0.333} < Q_{17-21\text{Sep}98} \leq \hat{\theta}_{17-21\text{Sep},0.667}\right] > 0.334$	35 $\hat{P}\left[T_{\text{DJF}98} > \hat{\tau}_{\text{DJF},0.667}\right] = 0.333$
12	$\hat{P}\left[Q_{17-21\text{Sep}98} > \hat{\theta}_{17-21\text{Sep},0.667}\right] \leq 0.333$	36 $\hat{P}\left[Q_{\text{DJF}98} \leq \hat{\theta}_{\text{DJF},0.333}\right] = 0.223$
13	$\hat{P}\left[T_{\text{Sep}98} \leq \hat{\tau}_{\text{Sep},0.333}\right] = 0.343$	37 $\hat{P}\left[Q_{\text{DJF}98} > \hat{\theta}_{\text{DJF},0.667}\right] = 0.443$
14	$\hat{P}\left[T_{\text{Sep}98} > \hat{\tau}_{\text{Sep},0.667}\right] = 0.323$	38 $\hat{P}\left[T_{\text{JFM}99} \leq \hat{\tau}_{\text{JFM},0.333}\right] = 0.443$
15	$\hat{P}\left[T_{\text{Sep}98} \leq \hat{\tau}_{\text{Sep},0.333}\right] > 0.333$	39 $\hat{P}\left[T_{\text{JFM}99} > \hat{\tau}_{\text{JFM},0.667}\right] = 0.223$
16	$\hat{P}\left[T_{\text{Sep}98} > \hat{\tau}_{\text{Sep},0.667}\right] \leq 0.333$	40 $\hat{P}\left[Q_{\text{JFM}99} \leq \hat{\theta}_{\text{JFM},0.333}\right] = 0.233$
17	$\hat{P}\left[Q_{\text{Sep}98} \leq \hat{\theta}_{\text{Sep},0.333}\right] = 0.333$	41 $\hat{P}\left[Q_{\text{JFM}99} > \hat{\theta}_{\text{JFM},0.667}\right] = 0.433$
18	$\hat{P}\left[Q_{\text{Sep}98} > \hat{\theta}_{\text{Sep},0.667}\right] = 0.333$	
19	$\hat{P}\left[T_{\text{SON}98} \leq \hat{\tau}_{\text{SON},0.333}\right] = 0.383$	
20	$\hat{P}\left[T_{\text{SON}98} > \hat{\tau}_{\text{SON},0.667}\right] = 0.283$	
21	$\hat{P}\left[Q_{\text{SON}98} \leq \hat{\theta}_{\text{SON},0.333}\right] = 0.283$	
22	$\hat{P}\left[Q_{\text{SON}98} > \hat{\theta}_{\text{SON},0.667}\right] = 0.383$	
23	$\hat{P}\left[Q_{\text{SON}98} \leq \hat{\theta}_{\text{SON},0.333}\right] \leq 0.333$	
24	$\hat{P}\left[\hat{\theta}_{\text{SON},0.333} < Q_{\text{SON}98} \leq \hat{\theta}_{\text{SON},0.667}\right] \leq 0.334$	

Figure 8-4. Selected outlook probability settings in priority order.

happens because of numerical error growth or round-off error in the solution algorithms. (Note that the fifth equation of the EC seasonal outlook also is not in conflict with the corresponding seventh and eighth equations of the NOAA 1- and 3-month outlook in Figure 8-3; its removal does not change the solution.)

Therefore, the 41 equations that are not shaded in Figure 8-3 are selected and used in the priority order indicated in Figure 8-4, to create a hydrology outlook for the Lake Superior basin beginning September 15, 1998. The priority order for this example is chronological (by end date), with air temperature before precipitation and then United States before Canada for each time period. Other priority orders are possible, of course, and are discussed subsequently. Note that some of the probability statements to be used cover periods starting prior to September 15 but the derivative outlook begins September 15. This is another example of using meteorology outlooks beginning earlier than derivative outlooks. Per the discussion in Chapter 6, these probability statements will actually be imperfectly satisfied.

These 41 outlook settings are used with inspection of the forty-seven 12.5-month time series, beginning September 15 from the available historical record of 1948–94 in Tables 7-1 and 7-2, and from the quantile estimates in Tables 7-3 and 7-4, to construct the equations in Figure 8-5. The shaded area in Figure 8-5 contains coefficients for the "slack variables," introduced and discussed in the previous sidebar to change consideration of the inequality constraints in Equations (8-11), such as those in Figure 8-4, into consideration of equality constraints in the procedural algorithms. The slack variable values are not of interest here, even though their solution must be pursued jointly with the other weights; see Exercise A2-8 in Appendix 2. [Outlooks 1 through 14, 17 through 22, and 26 through 41 in Figure 8-4 are defined in terms of a reference period of 1961–90 (see Chapter 4) and refer to quantile estimates in Tables 7-3 and 7-4 (also defined for 1961–90) to construct the corresponding equations in Figure 8-5. The reader can only manually check the corresponding equations in Figure 8-5 for these outlooks, by using Tables 7-3 and 7-4 to estimate the reference quantiles. Outlooks 15, 16, 23, 24, and 25 in Figure 8-4 are actually defined in terms of a reference period of 1963–93 (also see Chapter 4), and their reference quantile estimates cannot be checked from Tables 7-3 and 7-4. However, they are computed correctly with the software described in Appendix 2 in Exercise A2-8.]

The first row in the left-most matrix in Figure 8-5 corresponds to Equation (6-3) in which all weights sum to the number of scenarios (47, in this example). Rows 2 through 42 in Figure 8-5 correspond to the 41 equations and inequalities in Figure 8-4 with 13 slack variables ($w_{48}, \ldots , w_{60}$) added to convert the inequalities into equations.

Table 8-1 presents the solution of these equations, found by minimizing the deviation of weights from unity, as in Equations (8-11), by utilizing as many climate outlook settings as possible (procedural algorithm 2). All computations are with probabilities (both reference quantiles and forecasts) significant to three digits after the decimal point. Note from Table 8-1 that some weights are zeros, indicating that some historical scenarios are not used. However, all but the last 14 of the climatic outlooks in Figure 8-4 are used. By using this set of weights, the probabilistic hydrology outlooks for Lake Superior net basin supply can be built from Table 7-12 as in Table 8-2. Comparing Tables 8-2 and 7-14 makes visible the effect of using different meteorology outlooks for the (same) hydrology outlook. See Exercise A2-8 in Appendix 2 for this example, worked with the software described there.

ELIMINATING DIFFICULT EQUATIONS

Eliminating infeasible equations according to a priority ordering is essential in searching for a set of weights by using the optimization of Equations (8-11) or any search algorithm. Some priority orders may give satisfaction of more equations than others (Croley 1996, 1997b). For example, if the 41 equations identified in Figure 8-4 are used in the priority of their appearance there, as was done in the preceding example, then procedural algorithm 2 gives weights that satisfy the first 27 of those equations, when used with the 47 scenarios of the available historical Lake Superior basin meteorology record. Alternatively, if the priorities of these 41 equations are reversed, then equations 1 through 4, 6, 7, 10, 11 and 13 through 21 in Figure 8-4 are unused. Elimination of other than lowest-priority equations would lead to alternative solutions, too.

$$
\begin{bmatrix}
1111111111111111111111111111111111111111111111 & 0000000000000 \\
0000011010011100110000100111000010010010000010 & 0000000000000 \\
1001100101100011001111000000100001000001100110 1 & 0000000000000 \\
1000001000111000100011010101110001000000100000 & 0000000000000 \\
0101110111000100010000001010001101000100011111 & 0000000000000 \\
0000000010000010000000001000000100000000001000 & 1000000000000 \\
0000000000000000000000010010000001110000001001 0 & 0100000000000 \\
0111111001111001111001001001111100000010000010 0 & 0010000000000 \\
0111111011111011111011011111111111111011001111 10 & 0001000000000 \\
0000000000000000000001000000000000000001000100000 & 0000100000000 \\
1000110000101110001101000010001101001000000011 & 0000010000000 \\
1000111101111110011101101010100111110111011010011 & 0000001000000 \\
0000001101010000010000101000010010010101010000 & 0000000100000 \\
0101000010000010110000001011000011001010000010 & 0000000000000 \\
1000100000111100001110010000001000010001110000 1 & 0000000000000 \\
1010111101111101001111110100111100110101111110 1 & 0000000010000 \\
1000100000111100001111100000010000100011100001 & 0000000001000 \\
1010100010001001001101000000101101000001010000 1 & 0000000000000 \\
0001000001010100110010100000010010110100000110 0 & 0000000000000 \\
0001000000010000110000010101001100001100101110 & 0000000000000 \\
1000011010101111000010110100000100011000000001 & 0000000000000 \\
1000111010000011100010010011001010010000000100001 & 0000000000000 \\
0101000100010100100001100000100001101001011000 & 0000000000000 \\
1000111010000011000101001100101001000000010000 1 & 0000000000100 \\
0010000001101000101010000011000110001011000011 0 & 0000000000010 \\
1010111011101011101111001111101111001011010100111 & 0000000000001 \\
0011000100110000101000001000101010010101001101010 & 0000000000000 \\
1000111011000111010000010010000001101001000001 & 0000000000000 \\
0000011101100001110000100111010001000001000000011 & 0000000000000 \\
0010000100000000011010110000000100110100100100 0 & 0000000000000 \\
0111000010100010100000110100101010001010001000 10 & 0000000000000 \\
1000111001010001010110000010000100100011100110 1 & 0000000000000 \\
0001101000011110000101001100100010000010001001 1 & 0000000000000 \\
1100000000000000010010110011010000010100100000 0 & 0000000000000 \\
0000000010100110101000001010011100100000000010 & 0000000000000 \\
1001111101010000101001000010000110100011001110 1 & 0000000000000 \\
0001101101111011010101001000010000000011100011111 & 0000000000000 \\
1100010000000000010001011000100001101100010000 00 & 0000000000000 \\
0100000000100010101001110100011001000000000010 & 0000000000000 \\
0001110101001001000000001010100010100110011110 1 & 0000000000000 \\
0001000111111010000101001100010000100110000111 0 & 0000000000000 \\
1110111000000000010000110011101001000001000000 & 0000000000000
\end{bmatrix}
\times
\begin{bmatrix}
w_1 \\
w_2 \\
\vdots \\
w_{47} \\
w_{48} \\
w_{49} \\
\vdots \\
w_{60}
\end{bmatrix}
=
\begin{bmatrix}
47.000 \\
7.191 \\
24.111 \\
18.471 \\
12.831 \\
4.700 \\
9.400 \\
18.800 \\
37.600 \\
4.700 \\
15.651 \\
31.302 \\
15.651 \\
16.121 \\
15.181 \\
31.349 \\
15.651 \\
15.651 \\
15.651 \\
18.001 \\
13.301 \\
13.301 \\
18.001 \\
15.651 \\
15.698 \\
31.349 \\
15.651 \\
15.651 \\
14.241 \\
17.061 \\
15.651 \\
15.651 \\
14.241 \\
17.061 \\
15.651 \\
15.651 \\
10.481 \\
20.821 \\
20.821 \\
10.481 \\
10.951 \\
20.351
\end{bmatrix}
$$

Figure 8-5. Alternative representation of Equation (6-3) and the equations in Figure 8-4.

Inspecting the equations is always a good idea, to avoid grouping any equations that may be difficult to satisfy simultaneously. Eliminating a difficult-to-satisfy equation that is of only marginal interest to the user may allow the satisfaction of more equations in procedural algorithm 1 or the use of more scenarios in algorithm 2. While there is no good example of this in Figure 8-4, consider some of the equations there. The NOAA 8–14 day outlook for air temperature, represented by the first two equations in Figure 8-4, is for a warm week in September. Likewise, the NOAA 6–10 day outlook for air

Table 8-1. Outlook weights maximizing use of mixed meteorology outlooks in Figure 8-5 for the Lake Superior supply outlook example.

Index, i (1)	Weight, w_i (2)	Index, i (3)	Weight, w_i (4)	Index, i (5)	Weight, w_i (6)
1	1.236938	17	0.500728	33	0.423563
2	1.123837	18	0.637977	34	3.139953
3	0.340688	19	1.679239	35	0.686021
4	1.872959	20	2.351983	36	0.115635
5	0	21	1.454609	37	2.659130
6	0	22	0.470003	38	2.361626
7	0	23	0	39	0.784653
8	0	24	2.213430	40	2.046122
9	0	25	0.213985	41	1.933896
10	0	26	0.650371	42	0.883765
11	0	27	0.821754	43	1.354481
12	1.265380	28	1.750284	44	1.674546
13	0	29	2.144328	45	1.896745
14	0	30	2.761213	46	0.240658
15	1.951520	31	0	47	0
16	0.258156	32	1.099824		

temperature, represented by equations 5 through 9 in Figure 8-4, is for continued warm weather over the next 5 days. However, the NOAA 1-month outlook for air temperature and the EC 1-month outlook for air temperature are for a cool September. While it is certainly possible to have a cooler than normal September with warmer than normal periods within, it may be that for data sets other than the Lake Superior basin data used here, it has never occurred in the historical record. Then it would not be possible to find any meteorology time series segments that meet all of these conditions, and the user would have to eliminate some of these conditions.

Table 8-2. September 15, 1998, Lake Superior probabilistic outlook of net basin supply (mm) using mixed meteorology outlooks.

Month	Nonexceedance quantiles								
	3%	10%	20%	30%	50%	70%	80%	90%	97%
(1)	(2)	(3)	(4)	(5)	(6)	(7)	(8)	(9)	(10)
Sep 98	−20.1	−14.0	−1.96	9.46	30.4	45.9	52.3	62.8	89.8
Oct 98	−54.9	−20.1	3.98	14.0	26.0	54.1	56.7	72.5	97.4
Nov 98	−72.8	−50.3	−25.1	−22.6	−14.0	23.8	36.2	49.0	62.2
Dec 98	−73.4	−59.7	−50.3	−42.2	−34.6	−18.5	−15.6	0.66	4.29
Jan 99	−80.4	−63.4	−52.0	−48.0	−35.5	−30.7	−21.3	−6.75	−1.95
Feb 99	−47.3	−43.4	−33.5	−32.4	−25.0	−10.3	−7.49	2.10	21.3
Mar 99	−31.9	−8.71	−3.68	2.09	20.1	32.6	50.7	63.9	106
Apr 99	54.2	69.6	87.9	95.2	111	134	146	156	165
May 99	97.2	114	120	125	157	176	178	193	233
Jun 99	105	111	119	131	149	164	170	186	202
Jul 99	81.4	94.9	99.7	113	133	156	175	196	206
Aug 99	15.2	50.5	76.2	85.8	103	114	126	137	165
Sep 99	−15.6	13.7	34.2	50.7	60.0	105	122	150	167

Of course, it is also important to eliminate truly incompatible equations, as indicated in the first block of the algorithm of Figure 8-2 and as was done in the example of Figure 8-3. As noted previously, some of the shaded equations in Figure 8-3 are incompatible with the NOAA 1- and 3-month Climate Outlook probabilities. In particular, the following equations are incompatible:

$$\hat{P}\left[T_{SON98} \leq \hat{t}_{SON, 0.333}\right] = 0.383 \quad (\text{NOAA event probability}) \quad (8\text{-}20a)$$

$$\hat{P}\left[T_{SON98} > \hat{t}_{SON, 0.667}\right] = 0.283 \quad (\text{NOAA event probability}) \quad (8\text{-}20b)$$

$$\hat{P}\left[T_{SON98} \leq \hat{t}_{SON, 0.333}\right] \leq 0.333 \quad (\text{EC most probable event}) \quad (8\text{-}20c)$$

$$\hat{P}\left[\hat{t}_{SON, 0.333} < T_{SON98} \leq \hat{t}_{SON, 0.667}\right] > 0.334 \quad (\text{EC most probable event}) \quad (8\text{-}20d)$$

$$\hat{P}\left[T_{SON98} > \hat{t}_{SON, 0.667}\right] \leq 0.333 \quad (\text{EC most probable event}) \quad (8\text{-}20e)$$

Equations (8-20a) and (8-20c) are truly incompatible. Also, Equations (8-20a) and (8-20b) imply that the probability in Equation (8-20d) must be 0.334, which is truly incompatible with that equation. [Equation (8-20e) is not incompatible with the rest, but it is redundant with Equation (8-20b).]

Note that *any* meteorology probability forecast can be incorporated into a hydrology outlook because the forecast equations must be of one of the following general forms:

$$\hat{P}\left[z_1 < X \leq z_2\right] = a \quad (8\text{-}21a)$$

$$\hat{P}\left[z_1 < X \leq z_2\right] > a \quad (8\text{-}21b)$$

$$\hat{P}\left[z_1 < X \leq z_2\right] < a \quad (8\text{-}21c)$$

$$\hat{P}\left[z_1 < X \leq z_2\right] \geq a \quad (8\text{-}21d)$$

$$\hat{P}\left[z_1 < X \leq z_2\right] \leq a \quad (8\text{-}21e)$$

and all of these forms, or their converses, are considered in the development of Equations (8-11). Of course, if the user adds additional probability equations, he or she must also check for incompatibilities within the entire set.

EXTENSIONS
An artifact of the methodology diagrammed in Figure 8-2 is that eliminating a small number of equations from a previous solution (i.e., expanding the constraint space of the optimization) could result in the algorithm's eliminating even more equations in its new search for a solution (i.e., a search for an optimum solution to the expanded constraint space). That is, if there is an optimum solution to a set of equations such as those represented in Equations (8-15) that also satisfies Equations (8-18) or (8-19), it should theoretically be possible to eliminate some equations in Equations (8-15), redo the optimization, and then find that the new optimum solution no longer satisfies Equations (8-18) or (8-19). In this case, continuing with the procedure of Figure 8-2 would eliminate the lowest-priority equation and solve the optimization again until an optimum solution is identified that *does* satisfy Equations (8-18) or (8-19). Because the solution to a set of equations also satisfies any subset of those equations, one might be surprised that the procedure of Figure 8-2 could possibly continue searching for a solution by eliminating even more equations, after the constraint space had been broadened by removing some equations. While unobserved so far, this should be possible. It would result because the

algorithm only finds an optimum solution, which might not satisfy Equations (8-18) or (8-19), and disregards other feasible solutions to the equations. There are always other feasible solutions (combinations of weights that satisfy all the equations) that are not optimum. Those solutions are interesting too; the optimization in Figure 8-2 is only a device to find *a* solution that might also satisfy Equations (8-18) or (8-19). Unfortunately, systematic searches of the equations represented in Equations (8-15) for feasible solutions (not necessarily optimum) that also satisfy Equations (8-18) or (8-19) involve evaluation of numerous roots and are thus computationally impractical. If an acceptance criterion could be formulated for usable solutions (not necessarily optimum), then evaluation of all solutions would be unnecessary and a partial search algorithm might be built that is practical. Again, however, there is not an obvious way to guarantee that the length of the resulting search would be acceptably short.

Formulating an optimization problem allows for a general approach in determining operational hydrology weights in the face of multiple outlooks, where many solutions are possible but difficult to systematically evaluate. In the absence of a partial search algorithm for finding and evaluating other than optimum solutions (feasible weight combinations), one could modify the optimization objective function—for example, by replacing Equation (8-15a) for procedural algorithm 1 with:

$$\min\left[\sum_{i=1}^{n}(w_i - 1)^2 + \sum_{i=1}^{n+p} f(w_i) + \sum_{i=n+p+1}^{n+p+q} g(w_i) \right] \quad (8\text{-}22)$$

or for procedural algorithm 2 with:

$$\min\left[\sum_{i=1}^{n}(w_i - 1)^2 + \sum_{i=1}^{n} g(w_i) + \sum_{i=n+1}^{n+p} f(w_i) + \sum_{i=n+p+1}^{n+p+q} g(w_i) \right] \quad (8\text{-}23)$$

where

$$f(w) = M \quad w \le 0 \quad (8\text{-}24\text{a})$$
$$f(w) = 0 \quad w > 0 \quad (8\text{-}24\text{b})$$

$$g(w) = M \quad w < 0 \quad (8\text{-}24\text{c})$$
$$g(w) = 0 \quad w \ge 0 \quad (8\text{-}24\text{d})$$

and M is a very large number. Minimization would force positive or nonnegative solutions if they exist. However, these formulations are not amenable to the techniques employed here with respect to defining a Lagrangian function that is continuous and allowing linear equations that are solvable via the Gauss-Jordan method of elimination.

The use of different meteorology outlooks, defined over different spatial extents as well as different temporal extents, is a simple extension of the methodology and is the subject of the next chapter. Derived probabilistic hydrology outlooks may be sensitive (1) to the choice of the objective function used in the optimization, (2) to the priority order assigned by the user to the probabilistic meteorology outlooks, and (3) to the meteorology probability values interpreted by the user from agency outlooks. The effect of the first is best studied with additional research and is the subject of Chapter 10. Users in their own applications, however, may assess the effect of the latter two, by simply repeating all calculations with alternative priority assignments or probability values. A recomputation of weights and their application to create probabilistic hydrology outlooks does not require recreating the hydrology scenarios to which the weights are to be applied. This issue is further addressed in Chapter 11.

Chapter 9

SIMULTANEOUS SPATIAL OUTLOOKS

DERIVATIVE-FORECAST METHODOLOGY

Figure 9-1 diagrams a methodology that integrates modeling and near real-time data handling to support the operational hydrology approach for making probabilistic derivative outlooks. Block 1 in Figure 9-1 is the step, of preparing a model, that has to be carried out before an operational hydrology outlook procedure can be set up. Blocks 2 through 4 are steps required whenever the historical meteorology record (containing data that have gone through a final quality control) is updated. New, recently obtained data that have not yet gone through final quality control are referred to as "provisional data" in Figure 9-1. Provisional data include data obtained in real time or near real time and, whenever available, are used in blocks 5 and 6 to define the initial conditions for use in derivative models. Block 7 represents the actual generation of a derivative scenario corresponding to one of the meteorology scenarios used in the operational hydrology approach. It is repeated for all such meteorology scenarios available, and the results are used in block 8 to build the sample of all scenarios for use in blocks 9 and 10, where the sample is restructured to match a probabilistic meteorology outlook and used to estimate the probabilistic derivative outlook. Blocks 9 and 10 may be repeated for other probabilistic meteorology outlooks without requiring the modeling simulations each time (as noted at the end of Chapter 6).

This methodology is well suited to any application that is defined over an area small enough that expressions of probabilistic meteorology outlooks at one geographic point can represent the whole area. All examples used in this book to this point have been such applications: that is, each meteorology outlook map for a single event probability or most-probable event presented thus far could be understood as sufficient to represent the application area (the Lake Superior basin has been used to this point). Thus the expressions of multiple probabilistic meteorology outlooks [in, for example, Figures 4-5, 4-14, 8-3, and 8-4 or in Equations (4-4) through (4-13), (7-12), or (7-13)] all represent values at a point but are deemed to be applicable to an entire geographic area of interest.

Of course, application areas may also be so large that it may be inappropriate to use only a single spatial value for each meteorology outlook. An example in Great Lakes forecasting occurs in estimating probabilistic lake level outlooks. Figure 9-2 is a map of the Great Lakes and their respective basins. Lake levels are functions of (simultaneous) hydrology variables on all Great Lakes, because the lakes are connected and jointly regulated. However, a separate set of meteorology outlooks applies to each lake basin area, and the sets differ for different areas. The available connecting-channel routing and lake-regulation models require data on water supplies on all lakes simultaneously and can determine levels and outflows only *jointly* on all lakes, because all levels and outflows are interdependent. In a deterministic forecast, the forecast water supply scenarios for each and all of the Great Lakes can be used as inputs to these models to determine the (simultaneous) lake level forecast time series on each lake.

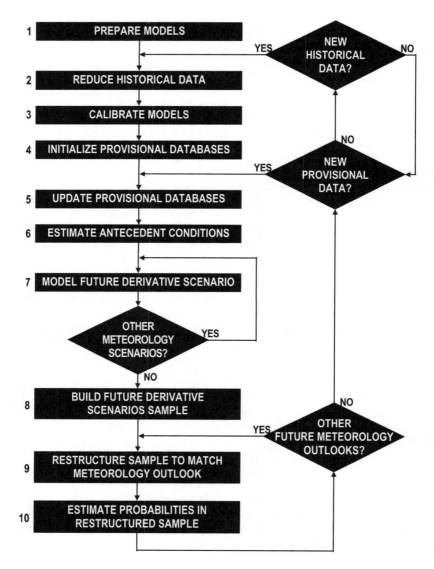

Figure 9-1. Derivative forecast methodology.

Direct application of this technique to a probabilistic outlook is not suitable. There is not a simple one-to-one transform between quantiles of water supply and lake levels. That is, the water supply that is exceeded 95% of the time (say) is not necessarily the amount corresponding to the water *level* that is exceeded 95% of the time. One *cannot* simply take the 95th percentile exceedance time series for water supplies on each lake as input to the connecting-channel routing and lake-regulation models to determine the 95th percentile exceedance time series for lake levels or outflows. It is more appropriate to use the entire generated sample of water supply scenarios to create a sample of lake level scenarios from which to generate probability outlooks. Figure 9-3 diagrams a

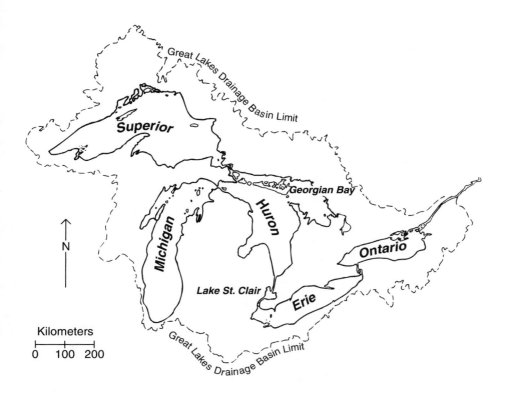

Figure 9-2. Great Lakes location map.

methodology appropriate for situations like these where derivative variables of interest include some defined only over each component area and some defined over the entire application area (these latter are functions of those variables defined over the component areas). The Great Lakes Environmental Research Laboratory (GLERL) employs this methodology to build a sample of scenarios for hydrology variables on each and all of the Great Lakes and of scenarios for consistent lake levels on all lakes. Blocks 7 and 8 in Figure 9-1 are replaced in Figure 9-3 to allow modeling of each derivative scenario, from each meteorology scenario, on all areas to proceed before the corresponding scenario over the entire area (lake levels in this case) is generated with other models. This is repeated for all meteorology scenarios to build the sample of consistent derivative scenarios.

Blocks 9 and 10 in Figure 9-3 require a single set of weights. However, while it is appropriate to represent each meteorology outlook over a single lake basin with a single set of equations [such as is given, for example, in Figure 8-4 by the first and second equations to represent the National Oceanic and Atmospheric Administration (NOAA) 8–14 day temperature outlook], it is not appropriate to represent each outlook with a single set of equations over the entire Great Lakes area at once. That means that a separate set of probabilistic meteorology outlooks must be used for each application area (lake basin). Therefore, because each application area involves a (generally) different set of probabilistic meteorology outlooks (and resulting set of weights, when calculated independently for each separate area), the resulting restructured (weighted) sample of sce-

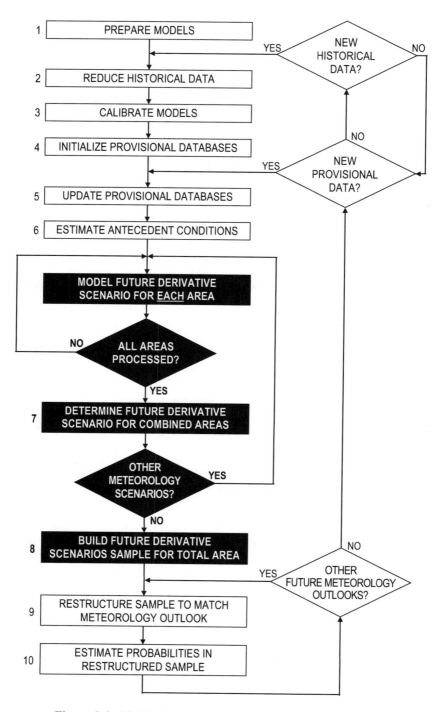

Figure 9-3. Multiple-area derivative forecast methodology.

narios (water supply scenarios in particular) does not correspond from lake to lake, and the scenarios would not be representative of the same sample when derived independently.

Therefore, a method is required to use and mix multiple probabilistic meteorology outlooks (both event probabilities and most-probable events), not only for different periods, lags, and variables, but also for different spatial extents. This includes, for example, probabilistic meteorology outlooks for warmer weather over Lake Superior *and* colder weather over Lake Michigan.

SIMULTANEOUS SPATIAL OUTLOOKS

As discussed earlier, the calculation of weights solves a set of equations representing multiple meteorology outlooks for a single application area. The goal is now to achieve a single set of weights for use in multiple-area applications, as in blocks 9 and 10 in Figure 9-3, from (different) simultaneous meteorology outlooks defined over the multiple areas. The solution is simply to first determine the sample of modeled derivative scenarios on all (component) application areas from corresponding sets of meteorology time series segments (historical record segments, in the examples presented in this book). Of course, the meteorology time series segments are defined over each application area, and they differ from one another but must correspond temporally. (In using historical record segments as the meteorology time series segments, this means using the same periods for all application areas.) This is illustrated in blocks 7 and 8 in both Figures 9-1 and 9-3. The next step is to derive an equation set for each application area from the relevant probabilistic meteorology outlooks over that area. *Then place all of these sets of equations together into one master set.* Finally, solve the master set of equations (representing all probabilistic meteorology outlooks over all application areas) *simultaneously* to determine one set of weights, as before. This satisfies all of the multiple spatial meteorology outlooks over all of the application areas. (It, of course, requires that forecast parameters, such as start date and length of forecast, and historical meteorology record periods are the same over all application areas.) The resulting weights can then be used directly to estimate probabilities for all variables, including those that are interrelated among the different application areas (such as Great Lakes water levels, in the present example).

Consider the following example for the multiple application areas of Lakes Superior, Michigan, Huron, and Erie, and Georgian Bay; see Figure 9-2. In creating a hydrology outlook on September 15, 1998, for extended Great Lakes level-pool water levels, GLERL wants to use NOAA's 3-month probabilistic outlooks of air temperature and precipitation over each lake basin for the September-October-November period. ("Level-pool" water levels ignore short-term fluctuations such as wind- or storm-generated waves and seiches.) This is another example of using a meteorology outlook that begins earlier than the derivative outlook. Per the discussion in Chapter 6, the corresponding probability statements will be imperfectly satisfied with actual, as opposed to historical, data. Relevant values are abstracted from Figures 4-3 and 4-4 and presented in Figure 9-4 in order of priority. Note in Figure 9-4 that an additional subscript has been added to the variables and quantiles to denote the application area ("Sup," "Mic," "Hur," "Geo," and "Eri" to denote Superior, Michigan, Huron, Georgian Bay, and Erie, respectively). The priority order is by lake (Superior, Michigan, Huron, Georgian Bay, and Erie), with precipitation outlooks first for each lake. The outlooks generally call for above normal SON precipitation on all lake basins, below normal SON air temperature

$$1 \qquad \hat{P}\left[Q_{\text{Sup, SON98}} \leq \hat{\theta}_{\text{Sup, SON, 0.333}}\right] = 0.283$$

$$2 \qquad \hat{P}\left[Q_{\text{Sup, SON98}} > \hat{\theta}_{\text{Sup, SON, 0.667}}\right] = 0.383$$

$$3 \qquad \hat{P}\left[T_{\text{Sup, SON98}} \leq \hat{\tau}_{\text{Sup, SON, 0.333}}\right] = 0.383$$

$$4 \qquad \hat{P}\left[T_{\text{Sup, SON98}} > \hat{\tau}_{\text{Sup, SON, 0.667}}\right] = 0.283$$

$$5 \qquad \hat{P}\left[Q_{\text{Mic, SON98}} \leq \hat{\theta}_{\text{Mic, SON, 0.333}}\right] = 0.283$$

$$6 \qquad \hat{P}\left[Q_{\text{Mic, SON98}} > \hat{\theta}_{\text{Mic, SON, 0.667}}\right] = 0.383$$

$$7 \qquad \hat{P}\left[T_{\text{Mic, SON98}} \leq \hat{\tau}_{\text{Mic, SON, 0.333}}\right] = 0.363$$

$$8 \qquad \hat{P}\left[T_{\text{Mic, SON98}} > \hat{\tau}_{\text{Mic, SON, 0.667}}\right] = 0.303$$

$$9 \qquad \hat{P}\left[Q_{\text{Hur, SON98}} \leq \hat{\theta}_{\text{Hur, SON, 0.333}}\right] = 0.283$$

$$10 \qquad \hat{P}\left[Q_{\text{Hur, SON98}} > \hat{\theta}_{\text{Hur, SON, 0.667}}\right] = 0.383$$

$$11 \qquad \hat{P}\left[T_{\text{Hur, SON98}} \leq \hat{\tau}_{\text{Hur, SON, 0.333}}\right] = 0.333$$

$$12 \qquad \hat{P}\left[T_{\text{Hur, SON98}} > \hat{\tau}_{\text{Hur, SON, 0.667}}\right] = 0.333$$

$$13 \qquad \hat{P}\left[Q_{\text{Geo, SON98}} \leq \hat{\theta}_{\text{Geo, SON, 0.333}}\right] = 0.303$$

$$14 \qquad \hat{P}\left[Q_{\text{Geo, SON98}} > \hat{\theta}_{\text{Geo, SON, 0.667}}\right] = 0.363$$

$$15 \qquad \hat{P}\left[T_{\text{Geo, SON98}} \leq \hat{\tau}_{\text{Geo, SON, 0.333}}\right] = 0.333$$

$$16 \qquad \hat{P}\left[T_{\text{Geo, SON98}} > \hat{\tau}_{\text{Geo, SON, 0.667}}\right] = 0.333$$

$$17 \qquad \hat{P}\left[Q_{\text{Eri, SON98}} \leq \hat{\theta}_{\text{Eri, SON, 0.333}}\right] = 0.313$$

$$18 \qquad \hat{P}\left[Q_{\text{Eri, SON98}} > \hat{\theta}_{\text{Eri, SON, 0.667}}\right] = 0.353$$

$$19 \qquad \hat{P}\left[T_{\text{Eri, SON98}} \leq \hat{\tau}_{\text{Eri, SON, 0.333}}\right] = 0.333$$

$$20 \qquad \hat{P}\left[T_{\text{Eri, SON98}} > \hat{\tau}_{\text{Eri, SON, 0.667}}\right] = 0.333$$

Figure 9-4.   **NOAA's September-October-November 1998 probabilistic meteorology outlooks for the Upper Great Lakes.**

over the Lake Superior and Lake Michigan basins, and normal SON air temperature over the Lake Huron and Lake Erie basins and Georgian Bay.

While there are forty-seven 12.5 month time series in the historical record (1948–94) for Lakes Superior, Michigan, Huron, and Erie, there are only 41 (1954–94) for Georgian Bay, because of the unavailability of Canadian meteorology data there. Since a common period must be used in the computation of the joint weights, only the 41 years from 1954 through 1994 can be used for all the lakes. The 20 equations in Figure 9-4 are used together with inspection of the forty-one 12.5-month times series, beginning September 15 from the available historical record of 1954–94 and from quantile estimates, to construct the equations in Figure 9-5. (See Exercise A2-9 in Appendix 2.)

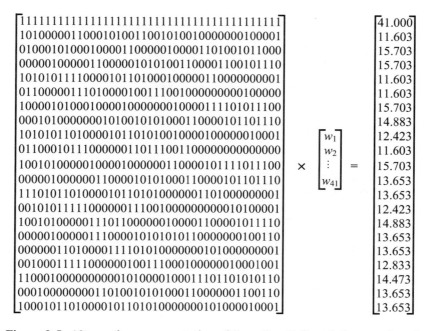

$$
\begin{bmatrix}
1111111111111111111111111111111111111111111 \\
1010000011000101001100101001000000100001 \\
0100010100010000110000010000110100101011000 \\
0000010000011000001010100110000110010110 \\
1010101111000010110100010000011000000000 01 \\
0110000011101000010011100100000000000100000 \\
10000101000100001000000010000111101011100 \\
0001010000000101001010100011000010110110 \\
101010110100001011010100100001000000100 01 \\
0110001011100000011011100110000000000000 \\
100101000001000010000001100001011110111 00 \\
0000010000001100001010100011000010110110 \\
111010110100001011010100000011010000000 01 \\
001010111110000001110010000000000010100001 \\
100101000001110110000001000011000010111 10 \\
0000010000011100001010101011000000010011 0 \\
000000110100001111010100000001010000000 01 \\
001000111110000001001110001000000010001001 \\
110001000000000010100001000111011010101 10 \\
000100000000110100101010001100000011001 10 \\
1000101101000010110101000000010100001000 1
\end{bmatrix}
\times
\begin{bmatrix}
w_1 \\
w_2 \\
\vdots \\
w_{41}
\end{bmatrix}
=
\begin{bmatrix}
41.000 \\
11.603 \\
15.703 \\
15.703 \\
11.603 \\
11.603 \\
15.703 \\
14.883 \\
12.423 \\
11.603 \\
15.703 \\
13.653 \\
13.653 \\
12.423 \\
14.883 \\
13.653 \\
13.653 \\
12.833 \\
14.473 \\
13.653 \\
13.653
\end{bmatrix}
$$

Figure 9-5. Alternative representation of Equation (6-3) and the equations in Figure 9-4.

Tables 7-1 and 7-2 contain SON values from the available historical record for the Lake Superior basin, and Tables 7-3 and 7-4 contain relevant quantile estimates for the Lake Superior basin. These values are repeated in Tables 9-1, 9-2, 9-3, and 9-4, respectively, which also present similar values for the other basin areas and selected composite areas. The first row in the matrix in Figure 9-5 corresponds to Equation (6-3) in which all weights sum to the number of scenarios (41 in this example). Rows 2 through 21 in Figure 9-5 correspond to the 20 equations in Figure 9-4. The reader can verify rows 2 through 5 in Figure 9-5 by using equations for the Lake Superior basin in Figure 9-4 and by inspecting Tables 9-1 through 9-4 (in all of which column 3 represents the Lake Superior basin). Similarly, rows 6 through 9 correspond to the Lake Michigan basin, 10 through 13 to the Lake Huron basin, 14 through 17 to the Georgian Bay basin, and 18 through 21 to the Lake Erie basin. (Columns 8 and 9 in Tables 9-1 and 9-3 and column 8 in Tables 9-2 and 9-4 correspond to composite areas, which are used in Chapter 10 examples.) Table 9-5 shows the solution of the equations in Figure 9-5, made with the second procedural algorithm (to maximize the number of meteorology outlooks used). (All computations are with probabilities, both reference quantiles and forecasts, significant to three digits after the decimal point.) Note in Table 9-5 that only two meteorology time series segments are unused (there are zero weights), but all of the meteorology outlooks are used for all application areas. On the basis of this set of weights, the probabilistic lake level outlooks for Lakes Superior, Michigan-Huron (including Georgian Bay), and Erie are estimated and presented in Table 9-6. Thus, even though different meteorology outlooks were used over the various lake basins, a simultaneous forecast of joint lake levels over all lakes results. See Exercise A2-9 in Appendix 2 for this example, worked with the software described there.

Table 9-1. Average SON air temperatures $\left(t_{\text{Area, SON}}\right)_i$ **(°C) for different areas.**

i	Year	Sup	Mic	Hur	Geo	Eri	Sup + Mic	Hur+Geo+Eri
(1)	(2)	(3)	(4)	(5)	(6)	(7)	(8)	(9)
1	1954	5.87	9.80	9.68	7.19	11.75	7.64	9.42
2	1955	5.50	9.23	9.20	6.89	11.01	7.19	8.92
3	1956	5.91	9.85	9.22	6.81	11.21	7.69	8.96
4	1957	5.52	8.71	8.71	7.06	10.51	6.96	8.67
5	1958	6.48	10.13	9.70	7.52	11.52	8.13	9.47
6	1959	3.74	8.10	8.40	6.19	11.05	5.71	8.42
7	1960	6.50	10.30	9.90	7.99	11.96	8.21	9.84
8	1961	6.55	10.41	10.64	8.64	12.61	8.30	10.53
9	1962	6.27	9.37	8.90	6.94	10.64	7.67	8.73
10	1963	8.40	11.66	11.04	8.78	12.46	9.87	10.66
11	1964	5.12	9.45	9.16	6.66	10.90	7.08	8.79
12	1965	4.61	9.22	8.84	6.23	11.09	6.69	8.59
13	1966	4.74	8.95	8.69	6.58	10.32	6.64	8.43
14	1967	5.04	8.23	7.99	5.99	9.52	6.48	7.74
15	1968	6.63	10.10	10.14	8.07	11.66	8.20	9.86
16	1969	5.15	8.75	9.00	8.37	10.52	6.77	9.24
17	1970	6.16	9.87	10.00	7.87	11.94	7.83	9.83
18	1971	7.04	11.20	11.04	8.99	12.77	8.92	10.83
19	1972	4.02	8.23	7.87	5.45	9.75	5.92	7.57
20	1973	6.47	10.56	10.15	7.77	12.35	8.32	9.97
21	1974	4.52	8.54	8.23	5.69	10.04	6.34	7.87
22	1975	5.75	9.92	9.93	7.67	11.37	7.63	9.56
23	1976	3.51	6.83	6.82	4.85	8.19	5.01	6.53
24	1977	5.84	9.09	8.93	7.11	11.00	7.31	8.91
25	1978	5.20	9.55	8.99	6.19	11.15	7.17	8.64
26	1979	4.67	8.94	9.01	6.99	10.86	6.60	8.85
27	1980	4.08	8.26	7.56	5.21	9.81	5.97	7.40
28	1981	5.41	8.41	7.85	5.92	9.94	6.77	7.80
29	1982	5.32	9.32	9.30	7.55	11.32	7.12	9.29
30	1983	6.51	9.83	9.44	7.60	11.55	8.01	9.43
31	1984	5.88	9.36	9.10	7.39	11.05	7.45	9.08
32	1985	4.60	9.12	9.38	7.59	12.03	6.64	9.55
33	1986	4.20	8.58	8.32	6.79	11.00	6.18	8.60
34	1987	5.75	8.99	8.76	6.65	10.61	7.21	8.57
35	1988	5.50	8.55	8.42	7.02	10.02	6.87	8.41
36	1989	4.91	8.34	8.14	6.38	10.34	6.46	8.18
37	1990	5.77	9.77	9.16	7.15	11.54	7.58	9.17
38	1991	4.05	8.11	8.43	6.92	10.76	5.89	8.60
39	1992	4.52	7.97	7.86	6.08	10.13	6.08	7.91
40	1993	3.16	7.49	7.37	5.29	9.76	5.11	7.36
41	1994	7.50	10.58	10.13	8.32	11.91	8.89	10.03

Table 9-2. Total SON precipitation $\left(q_{\text{Area, SON}}\right)_i$ **(mm) for different areas.**

i (1)	Year (2)	Sup (3)	Mic (4)	Hur (5)	Geo (6)	Eri (7)	Sup+Mic+Hur+Geo+Eri (8)
1	1954	200.20	290.29	326.69	346.71	300.30	276.64
2	1955	278.46	188.37	186.55	256.62	273.00	238.42
3	1956	172.90	110.11	131.04	216.58	122.85	150.15
4	1957	239.33	210.21	257.53	333.97	244.79	249.34
5	1958	227.50	208.39	224.77	242.97	241.15	225.68
6	1959	272.09	294.84	289.38	332.15	283.92	290.29
7	1960	216.58	204.75	174.72	229.32	126.49	195.65
8	1961	250.25	303.03	222.04	242.97	195.65	250.25
9	1962	154.70	156.52	192.01	193.83	218.40	174.72
10	1963	148.33	154.70	155.61	171.99	114.66	148.33
11	1964	226.59	194.74	178.36	241.15	103.74	194.74
12	1965	304.85	311.22	266.63	338.52	239.33	295.75
13	1966	209.30	197.47	236.60	321.23	232.05	229.32
14	1967	182.00	221.13	235.69	299.39	243.88	225.68
15	1968	241.15	230.23	226.59	255.71	232.05	236.60
16	1969	195.65	224.77	239.33	313.04	232.05	230.23
17	1970	312.13	313.95	267.54	300.30	275.73	298.48
18	1971	286.65	194.74	149.24	193.83	183.82	213.85
19	1972	192.92	245.70	192.01	193.83	308.49	222.04
20	1973	201.11	205.66	205.66	207.48	219.31	206.57
21	1974	217.49	177.45	193.83	272.09	208.39	210.21
22	1975	242.06	183.82	177.45	251.16	197.47	212.03
23	1976	122.85	108.29	169.26	223.86	200.20	151.97
24	1977	283.01	253.89	277.55	307.58	298.48	281.19
25	1978	191.10	262.99	258.44	281.19	219.31	235.69
26	1979	244.79	182.00	201.11	276.64	261.17	229.32
27	1980	241.15	202.93	202.93	254.80	199.29	221.13
28	1981	177.45	226.59	256.62	258.44	279.37	229.32
29	1982	300.30	249.34	236.60	321.23	279.37	278.46
30	1983	317.59	280.28	273.00	307.58	298.48	296.66
31	1984	220.22	293.93	254.80	295.75	225.68	256.62
32	1985	349.44	348.53	308.49	290.29	343.98	333.97
33	1986	237.51	334.88	362.18	269.36	306.67	293.93
34	1987	209.30	234.78	264.81	235.69	213.85	229.32
35	1988	287.56	343.98	335.79	354.90	282.10	318.50
36	1989	182.00	170.17	214.76	249.34	241.15	202.02
37	1990	251.16	286.65	313.04	347.62	300.30	291.20
38	1991	321.23	328.51	258.44	317.59	217.49	297.57
39	1992	242.06	312.13	289.38	305.76	340.34	290.29
40	1993	218.40	222.04	229.32	314.86	275.73	242.06
41	1994	200.20	218.40	210.21	233.87	171.99	207.48

Table 9-3 Ordered average SON air temperatures 1961–90

$$\left(t_{\text{Area, SON}}\right)_i \text{ (°C) for different areas.}$$

i	Sup	Mic	Hur	Geo	Eri	Sup+Mic	Hur+Geo+Eri
(1)	(3)	(4)	(5)	(6)	(7)	(8)	(9)
1	3.51	6.83	6.82	4.85	8.19	`5.01	6.53
2	4.02	8.23	7.56	5.21	9.52	5.92	7.40
3	4.08	8.23	7.85	5.45	9.75	5.97	7.57
4	4.20	8.26	7.87	5.69	9.81	6.18	7.74
5	4.52	8.34	7.99	5.92	9.94	6.34	7.80
6	4.60	8.41	8.14	5.99	10.02	6.46	7.87
7	4.61	8.54	8.23	6.19	10.04	6.48	8.18
8	4.67	8.55	8.32	6.23	10.32	6.60	8.41
9	4.74	8.58	8.42	6.38	10.34	6.64	8.43
10	4.91	8.75	8.69	6.58	10.52	6.64	8.57
11	5.04	8.94	8.76	6.65	10.61	6.69	8.59
12	5.12	8.95	8.84	6.66	10.64	6.77	8.60
13	5.15	8.99	8.90	6.79	10.86	6.77	8.64
14	5.20	9.09	8.93	6.94	10.90	6.87	8.73
15	5.32	9.12	8.99	6.99	11.00	7.08	8.79
16	5.41	9.22	9.00	7.02	11.00	7.12	8.85
17	5.50	9.32	9.01	7.11	11.05	7.17	8.91
18	5.75	9.36	9.10	7.15	11.09	7.21	9.08
19	5.75	9.37	9.16	7.39	11.15	7.31	9.17
20	5.77	9.45	9.16	7.55	11.32	7.45	9.24
21	5.84	9.55	9.30	7.59	11.37	7.58	9.29
22	5.88	9.77	9.38	7.60	11.54	7.63	9.43
23	6.16	9.83	9.44	7.67	11.55	7.67	9.55
24	6.27	9.87	9.93	7.77	11.66	7.83	9.56
25	6.47	9.92	10.00	7.87	11.94	8.01	9.83
26	6.51	10.10	10.14	8.07	12.03	8.20	9.86
27	6.55	10.41	10.15	8.37	12.35	8.30	9.97
28	6.63	10.56	10.64	8.64	12.46	8.32	10.53
29	7.04	11.20	11.04	8.78	12.61	8.92	10.66
30	8.40	11.66	11.04	8.99	12.77	9.87	10.83

Table 9-4. Ordered total SON precipitation 1961–90, $\left(q_{\text{Area, SON}}\right)_i$ (mm) for different areas.

i	i/n	Sup	Mic	Hur	Geo	Eri	Sup+Mic+Hur+Geo+Eri
(1)	(2)	(3)	(4)	(5)	(6)	(7)	(8)
1	0.033	122.85	108.29	149.24	171.99	103.74	148.33
2	0.067	148.33	154.70	155.61	193.83	114.66	151.97
3	0.100	154.70	156.52	169.26	193.83	183.82	174.72
4	0.133	177.45	170.17	177.45	193.83	195.65	194.74
5	0.167	182.00	177.45	178.36	207.48	197.47	202.02
6	0.200	182.00	182.00	192.01	223.86	199.29	206.57
7	0.233	191.10	183.82	192.01	235.69	200.20	210.21
8	0.267	192.92	194.74	193.83	241.15	208.39	212.03
9	0.300	195.65	194.74	201.11	242.97	213.85	213.85
10	0.333	201.11	197.47	202.93	249.34	218.40	221.13
11	0.367	209.30	202.93	205.66	251.16	219.31	222.04
12	0.400	209.30	205.66	214.76	254.80	219.31	225.68
13	0.433	217.49	221.13	222.04	255.71	225.68	229.32
14	0.467	220.22	224.77	226.59	258.44	232.05	229.32
15	0.500	226.59	226.59	235.69	269.36	232.05	229.32
16	0.533	237.51	230.23	236.60	272.09	232.05	229.32
17	0.567	241.15	234.78	236.60	276.64	239.33	230.23
18	0.600	241.15	245.70	239.33	281.19	241.15	235.69
19	0.633	242.06	249.34	254.80	290.29	243.88	236.60
20	0.667	244.79	253.89	256.62	295.75	261.17	250.25
21	0.700	250.25	262.99	258.44	299.39	275.73	256.62
22	0.733	251.16	280.28	264.81	300.30	279.37	278.46
23	0.767	283.01	286.65	266.63	307.58	279.37	281.19
24	0.800	286.65	293.93	267.54	307.58	282.10	291.20
25	0.833	287.56	303.03	273.00	313.04	298.48	293.93
26	0.867	300.30	311.22	277.55	321.23	298.48	295.75
27	0.900	304.85	313.95	308.49	321.23	300.30	296.66
28	0.933	312.13	334.88	313.04	338.52	306.67	298.48
29	0.967	317.59	343.98	335.79	347.62	308.49	318.50
30	1.000	349.44	348.53	362.18	354.90	343.98	333.97

Table 9-5. Outlook weights maximizing use of mixed spatial meteorology outlooks in Figure 9-4 for Superior, Michigan, Huron, Georgian Bay, and Erie application areas.

Index, i (1)	Weight, w_i (2)	Index, i (3)	Weight, w_i (4)	Index, i (5)	Weight, w_i (6)
1	0	15	1.041703	29	0.572903
2	0.984499	16	1.726796	30	0.815944
3	0	17	0.815944	31	1.312137
4	0.722075	18	1.254587	32	2.462463
5	0.169395	19	1.013244	33	0.765602
6	0.824203	20	0.998046	34	1.274970
7	0.512176	21	1.013744	35	1.277586
8	1.646760	22	1.024422	36	1.388151
9	0.720295	23	0.970088	37	1.557401
10	0.512124	24	0.961853	38	0.859067
11	1.121130	25	1.232463	39	0.463639
12	1.669792	26	1.395088	40	0.577446
13	1.218871	27	1.081603	41	0.842034
14	0.819483	28	1.380274		

Table 9-6. Upper Great Lakes probabilistic outlook using meteorology outlooks in Figure 9-4.

Month	Nonexceedance quantiles						
	10%	20%	30%	50%	70%	80%	90%
(1)	(2)	(3)	(4)	(5)	(6)	(7)	(8)
Lake Superior water level (m above mean sea level)							
Sep 98	183.32	183.32	183.32	183.33	183.33	183.33	183.34
Oct 98	183.26	183.28	183.29	183.31	183.32	183.33	183.36
Nov 98	183.22	183.23	183.25	183.27	183.29	183.33	183.38
Dec 98	183.13	183.16	183.18	183.20	183.23	183.28	183.33
Jan 99	183.03	183.07	183.09	183.12	183.17	183.19	183.24
Feb 99	182.96	183.00	183.01	183.04	183.09	183.10	183.16
Mar 99	182.90	182.94	182.95	182.99	183.03	183.07	183.12
Apr 99	182.92	182.94	182.96	183.01	183.08	183.10	183.13
May 99	183.03	183.06	183.06	183.08	183.16	183.19	183.22
Jun 99	183.14	183.15	183.17	183.21	183.26	183.27	183.32
Jul 99	183.20	183.22	183.26	183.30	183.35	183.37	183.41
Aug 99	183.25	183.28	183.31	183.35	183.40	183.43	183.47
Sep 99	183.28	183.32	183.33	183.38	183.42	183.45	183.52
Lake Michigan-Huron-Georgian Bay water level (m above mean sea level)							
Sep 98	176.67	176.68	176.68	176.69	176.69	176.70	176.71
Oct 98	176.54	176.56	176.59	176.60	176.62	176.63	176.65
Nov 98	176.45	176.47	176.50	176.53	176.57	176.60	176.61
Dec 98	176.35	176.38	176.41	176.46	176.50	176.53	176.57
Jan 99	176.23	176.29	176.31	176.38	176.43	176.46	176.49
Feb 99	176.19	176.23	176.26	176.31	176.40	176.41	176.44
Mar 99	176.15	176.23	176.25	176.30	176.40	176.42	176.45
Apr 99	176.23	176.26	176.30	176.36	176.46	176.49	176.51
May 99	176.30	176.35	176.40	176.46	176.56	176.56	176.62
Jun 99	176.32	176.44	176.46	176.52	176.62	176.65	176.71
Jul 99	176.34	176.45	176.51	176.55	176.67	176.69	176.71
Aug 99	176.33	176.45	176.50	176.55	176.66	176.68	176.70
Sep 99	176.29	176.38	176.47	176.52	176.60	176.62	176.72
Lake Erie water level (m above mean sea level)							
Sep 98	174.45	174.45	174.45	174.46	174.47	174.47	174.48
Oct 98	174.25	174.27	174.27	174.30	174.32	174.34	174.37
Nov 98	174.13	174.16	174.17	174.22	174.28	174.30	174.33
Dec 98	174.09	174.12	174.18	174.23	174.27	174.31	174.36
Jan 99	174.08	174.13	174.16	174.22	174.28	174.34	174.36
Feb 99	174.08	174.16	174.18	174.22	174.30	174.32	174.38
Mar 99	174.16	174.18	174.23	174.28	174.33	174.40	174.42
Apr 99	174.21	174.28	174.31	174.37	174.43	174.46	174.52
May 99	174.26	174.33	174.39	174.43	174.48	174.48	174.55
Jun 99	174.20	174.33	174.37	174.44	174.49	174.51	174.54
Jul 99	174.17	174.29	174.36	174.41	174.47	174.50	174.51
Aug 99	174.11	174.23	174.30	174.37	174.40	174.42	174.44
Sep 99	174.04	174.16	174.20	174.27	174.33	174.33	174.37

Chapter 10

IMPROVED DERIVATIVE OUTLOOKS

Chapters 6 and 7 discussed the use of multiple probabilistic meteorology outlooks in the form of event probabilities, and Chapter 8 discussed the mixing of multiple event probabilities and multiple most-probable events. The approach involves restructuring an operational hydrology sample of possible "futures" so that various probabilities, or probability relationships, for meteorology variables match agency outlooks. Because there are typically many ways to restructure the sample (many sets of weights could be derived to weight the sample), it is necessary to define a method for selecting among them. Chapters 7, 8, and 9 used the minimization of the sum of squared differences of each weight with unity as an objective in an optimization over values of weights defined by the equations representing the agency outlooks. That objective strives to find weights closest to unity, representing a minimum biasing of scenarios and giving derived forecast probabilities closest to the relative frequencies found in the historical record (a solution nearest to climatology). Probability inequalities (representing the most-probable event type of probabilistic meteorology outlook) are satisfied at or near their limits. For example, a meteorology outlook stating "the average September air temperature is expected to be in the lower third of historical observations" is interpreted as

$$\hat{P}\left[T_{\text{Sep}} \leq \hat{\tau}_{\text{Sep, 0.333}}\right] > 0.333 \tag{10-1a}$$

$$\hat{P}\left[\hat{\tau}_{\text{Sep, 0.333}} < T_{\text{Sep}} \leq \hat{\tau}_{\text{Sep, 0.667}}\right] \leq 0.334 \tag{10-1b}$$

$$\hat{P}\left[T_{\text{Sep}} > \hat{\tau}_{\text{Sep, 0.667}}\right] \leq 0.333 \tag{10-1c}$$

and the minimization of the sum of squared differences of each weight with unity may give values of weights that satisfy the inequalities in Equations (10-1) only marginally:

$$\hat{P}\left[T_{\text{Sep}} \leq \hat{\tau}_{\text{Sep, 0.333}}\right] = 0.333 + \varepsilon_1 \cong 0.333 \tag{10-2a}$$

$$\hat{P}\left[\hat{\tau}_{\text{Sep, 0.333}} < T_{\text{Sep}} \leq \hat{\tau}_{\text{Sep, 0.667}}\right] = 0.334 - \varepsilon_2 \cong 0.334 \tag{10-2b}$$

$$\hat{P}\left[T_{\text{Sep}} > \hat{\tau}_{\text{Sep, 0.667}}\right] = 0.333 - \varepsilon_3 \cong 0.333 \tag{10-2c}$$

where ε_1, ε_2, and ε_3 are very small positive numbers. Likewise, even though probability equalities (representing the event probability type of probabilistic meteorology outlooks) are satisfied exactly in the solution, the resulting weighted operational hydrology sample may be as close as possible to climatology in all other regards.

OBJECTIVE REFORMULATION

This behavior makes sample restructuring sometimes relatively insensitive to meteorology outlooks. Instead, it may be more appropriate to use objective functions that strive to increase the difference between the actual probabilities selected and the limits of the constraining probability inequalities, where there are inequalities (of course). For the example of Equations (10-1), the optimization problem could be formulated as

$$\max \hat{P}\left[T_{\text{Sep}} \leq \hat{\tau}_{\text{Sep}, 0.333}\right] \text{ subject to} \tag{10-3a}$$

$$\hat{P}\left[T_{\text{Sep}} \leq \hat{\tau}_{\text{Sep}, 0.333}\right] > 0.333 \tag{10-3b}$$

$$\hat{P}\left[\hat{\tau}_{\text{Sep}, 0.333} < T_{\text{Sep}} \leq \hat{\tau}_{\text{Sep}, 0.667}\right] \leq 0.334 \tag{10-3c}$$

$$\hat{P}\left[T_{\text{Sep}} > \hat{\tau}_{\text{Sep}, 0.667}\right] \leq 0.333 \tag{10-3d}$$

or as

$$\max \left\{\hat{P}\left[T_{\text{Sep}} \leq \hat{\tau}_{\text{Sep}, 0.333}\right] - \hat{P}\left[\hat{\tau}_{\text{Sep}, 0.333} < T_{\text{Sep}} \leq \hat{\tau}_{\text{Sep}, 0.667}\right]\right.$$
$$\left. - \hat{P}\left[T_{\text{Sep}} > \hat{\tau}_{\text{Sep}, 0.667}\right]\right\} \text{ subject to} \tag{10-4a}$$

$$\hat{P}\left[T_{\text{Sep}} \leq \hat{\tau}_{\text{Sep}, 0.333}\right] > 0.333 \tag{10-4b}$$

$$\hat{P}\left[\hat{\tau}_{\text{Sep}, 0.333} < T_{\text{Sep}} \leq \hat{\tau}_{\text{Sep}, 0.667}\right] \leq 0.334 \tag{10-4c}$$

$$\hat{P}\left[T_{\text{Sep}} > \hat{\tau}_{\text{Sep}, 0.667}\right] \leq 0.333 \tag{10-4d}$$

These formulations strive to give a more extreme solution satisfying Equations (10-1) than is given by Equations (10-2).

The objective function formulation can be used to express many different goals, besides the one represented in Equation (10-3a) or (10-4a) for giving extreme probabilities that match meteorology outlooks. For example, in a multiple spatial outlook (representing multiple application areas), it may be desired to minimize or maximize the probability of events defined over the entire spatial extent. Recall the example of Figure 9-4, which was for wet and cool weather for the Superior and Michigan basins and wet and average-temperature weather for the Huron, Georgian Bay, and Erie basins over September, October, and November (SON). An over-riding goal might be to maximize the estimated probability of total precipitation in the upper third of its range for the entire area, reflecting generally wet weather over the entire application area:

$$\max \hat{P}\left[Q_{\text{Sup+Mic+Hur+Geo+Eri, SON98}} > \hat{\theta}_{\text{Sup+Mic+Hur+Geo+Eri, SON, 0.667}}\right] \tag{10-5}$$

or to maximize the estimated joint probability of total precipitation exceeding the median for the entire area and of air temperature not exceeding the median for the Superior and Michigan basins:

$$\max \hat{P}\left[\begin{array}{c} Q_{\text{Sup+Mic+Hur+Geo+Eri, SON98}} > \hat{\theta}_{\text{Sup+Mic+Hur+Geo+Eri, SON, 0.5}} \\ \text{and } T_{\text{Sup+Mic, SON98}} \leq \hat{\tau}_{\text{Sup+Mic, SON, 0.5}} \end{array}\right] \tag{10-6}$$

or to *minimize* the estimated joint probability of total precipitation in the lower third of its range for the entire area and of air temperature in the upper third of its range for the Superior and Michigan basins and in both the lower and upper thirds of its range for the Huron, Georgian Bay, and Erie basins:

$$\min \hat{P}\left[\begin{array}{c} Q_{\text{Sup+Mic+Hur+Geo+Eri, SON98}} \leq \hat{\theta}_{\text{Sup+Mic+Hur+Geo+Eri, SON, 0.333}} \\ \text{and } T_{\text{Sup+Mic, SON98}} > \hat{\tau}_{\text{Sup+Mic, SON, 0.667}} \\ \text{and } T_{\text{Hur+Geo+Eri, SON98}} \leq \hat{\tau}_{\text{Hur+Geo+Eri, SON, 0.333}} \\ \text{and } T_{\text{Hur+Geo+Eri, SON98}} > \hat{\tau}_{\text{Hur+Geo+Eri, SON, 0.667}} \end{array}\right] \tag{10-7}$$

Another example would be to create an objective function expressing the goal of returning to near-climatic conditions beyond the near term by minimizing deviations from selected long-term normal meteorology probabilities.

The problem with the use of alternative objective functions is that the solution of the corresponding optimization problem may be intractable. However, if the objective function is always a statement of maximizing or minimizing a *probability*, as in the examples of Equations (10-3) through (10-7), then a reformulation of the entire problem can enable application of other standard optimization techniques.

ALTERNATIVE OPTIMIZATION
In particular, the objective function formulations of Equations (10-3) through (10-7) can be expressed in terms of weights by matching relative frequencies in the operational hydrology sample, as was done in Chapter 6 to replace Equation (6-1) with Equation (6-2) by using Equation (5-8), or as was done in Chapter 8 to replace Equations (8-6) with Equations (8-7) by using Equation (5-8). The reformulation of Equations (10-3) through (10-7) become, respectively

$$\max \left[\frac{1}{n} \sum_{i \mid (t_{Sep})_i \leq \hat{\tau}_{Sep,0.333}} w_i \right] \tag{10-8}$$

$$\max \left[\begin{array}{c} \frac{1}{n} \displaystyle\sum_{i \mid (t_{Sep})_i \leq \hat{\tau}_{Sep,0.333}} w_i \;-\; \frac{1}{n} \displaystyle\sum_{i \mid \hat{\tau}_{Sep,0.333} < (t_{Sep})_i \leq \hat{\tau}_{Sep,0.667}} w_i \\[2ex] -\; \frac{1}{n} \displaystyle\sum_{i \mid (t_{Sep})_i > \hat{\tau}_{Sep,0.667}} w_i \end{array} \right] \tag{10-9}$$

$$\max \left[\frac{1}{n} \sum_{i \mid (q_{Sup+Mic+Hur+Geo+Eri,SON})_i > \hat{\theta}_{Sup+Mic+Hur+Geo+Eri,SON,0.667}} w_i \right] \tag{10-10}$$

$$\max \left[\frac{1}{n} \sum_{\substack{i \mid (q_{Sup+Mic+Hur+Geo+Eri,SON})_i > \hat{\theta}_{Sup+Mic+Hur+Geo+Eri,SON,0.5} \\ \text{and } (t_{Sup+Mic,SON})_i \leq \hat{\tau}_{Sup+Mic,SON,0.5}}} w_i \right] \tag{10-11}$$

$$\max \left[\frac{1}{n} \sum_{\substack{i \mid \text{not} \left[\begin{array}{c} (q_{Sup+Mic+Hur+Geo+Eri,SON})_i \leq \hat{\theta}_{Sup+Mic+Hur+Geo+Eri,SON,0.333} \\ \text{and } (t_{Sup+Mic,SON})_i > \hat{\tau}_{Sup+Mic,SON,0.667} \\ \text{and } (t_{Hur+Geo+Eri,SON})_i \leq \hat{\tau}_{Hur+Geo+Eri,SON,0.333} \\ \text{and } (t_{Hur+Geo+Eri,SON})_i > \hat{\tau}_{Hur+Geo+Eri,SON,0.667} \end{array} \right]}} w_i \right] \tag{10-12}$$

Note in Equation (10-12), corresponding to Equation (10-7), that the following was used:

$$\max \hat{P}\left[A^c \right] = \max \left[1 - \hat{P}[A] \right] = 1 - \max \left[\hat{P}[A] \right] = 1 + \min \hat{P}[A] \tag{10-13}$$

Therefore the maximization of $\hat{P}\left[A^c\right]$ gives the same values of w_i, $i = 1, \ldots, n$, as does the minimization of $\hat{P}[A]$.

Each example, from Equations (10-8) through (10-12), can in turn be expressed in the general form

$$\max \sum_{i=1}^{n} \alpha_{0,i} w_i \qquad (10\text{-}14)$$

where the $\alpha_{0,i}$ are defined, as they were in Equations (6-10) and (8-8), as 1 or 0, corresponding to the inclusion or exclusion, of each variable in the respective appropriate sets of Equations (10-8) through (10-12). The generalized problem statement of Equations (8-10) can again be formulated as an optimization problem, as in Equations (8-11), but with the generalized probabilistic objective function of Equation (10-14)

$$\max \sum_{i=1}^{n} \alpha_{0,i} w_i \text{ subject to} \qquad (10\text{-}15a)$$

$$\sum_{i=1}^{n} \alpha_{k,i} w_i \;=\; e_k \qquad k = 1, \ldots, m \qquad (10\text{-}15b)$$

$$\sum_{i=1}^{n} \alpha_{k,i} w_i \;\leq\; e_k \qquad k = m+1, \ldots, m+u \qquad (10\text{-}15c)$$

$$w_i \;\geq\; 0 \qquad i = 1, \ldots, n \qquad (10\text{-}15d)$$

where there are m equality constraints (representing multiple event probability meteorology outlooks and the condition that the weights sum to n), u inequality constraints (representing multiple most-probable event meteorology outlooks), and n time series segments available in the operational hydrology sample. Note in the formulation of Equations (10-15) that the p strictly-less-than inequalities and the q less-than-or-equal-to inequalities in Equations (8-10) are replaced with u less-than-or-equal-to inequalities (i.e., $u = p + q$, and no distinction is made between the two types of inequalities). Also, in Equations (8-11) the implied positivity constraints $w_i > 0$ of procedural algorithm 1 or the implied nonnegativity constraints $w_i \geq 0$ of procedural algorithm 2 are now represented *explicitly*, but only with nonnegativity constraints in Equations (10-15). Equations (10-15c) and (10-15d) are not equivalent replacements to Equations (8-10b) and (8-10c) or to the implied positivity or nonnegativty constraints by any means. However, they represent an alternative formulation that is both amenable to specific optimization techniques and another way of viewing the problem.

LINEAR PROGRAMMING

Equations (10-15) are amenable to standard "linear programming" optimization techniques. An algebraic procedure, termed the *Simplex* method, has been developed (Hillier and Lieberman 1969, pp. 127-71) which progressively approaches the optimum solution through a well-defined iterative process until optimality is finally reached. Prior to application of the Simplex method, the equations and inequalities in Equations (10-15) are transformed into an equivalent two-stage problem. First, Equations (10-15) may be written all in terms of inequalities:

$$\max \sum_{i=1}^{n} \alpha_{0,i} w_i \text{ subject to} \tag{10-16a}$$

$$\sum_{i=1}^{n} \alpha_{k,i} w_i \leq e_k \qquad k = 1, \ldots, m \tag{10-16b}$$

$$\sum_{i=1}^{n} \alpha_{k,i} w_i \geq e_k \qquad k = 1, \ldots, m \tag{10-16c}$$

$$\sum_{i=1}^{n} \alpha_{k,i} w_i \leq e_k \qquad k = m+1, \ldots, m+u \tag{10-16d}$$

$$w_i \geq 0 \qquad i = 1, \ldots, n \tag{10-16e}$$

where the equalities in Equation (10-15b) have been replaced (equivalently) with two sets of inequalities in Equations (10-16b) and (10-16c). The solution to Equations (10-16) is identical to that of Equations (10-15). Furthermore, the greater-than-or-equal-to inequalities in Equation (10-16c) can be summed into a single equation without changing the solution:

$$\max \sum_{i=1}^{n} \alpha_{0,i} w_i \text{ subject to} \tag{10-17a}$$

$$\sum_{i=1}^{n} \alpha_{k,i} w_i \leq e_k \qquad k = 1, \ldots, m+u \tag{10-17b}$$

$$\sum_{k=1}^{m} \sum_{i=1}^{n} \alpha_{k,i} w_i \geq \sum_{k=1}^{m} e_k \tag{10-17c}$$

$$w_i \geq 0 \qquad i = 1, \ldots, n \tag{10-17d}$$

Then, Equations (10-17) may be written all in terms of equalities (except for the non-negativity constraints) by adding slack variables, as in Equations (8-12),

$$\max \sum_{i=1}^{n} \alpha_{0,i} w_i \text{ subject to} \tag{10-18a}$$

$$\sum_{i=1}^{n} \alpha_{k,i} w_i + w_{n+k} = e_k \qquad k = 1, \ldots, m+u \tag{10-18b}$$

$$\sum_{k=1}^{m} \sum_{i=1}^{n} \alpha_{k,i} w_i - w_{n+m+u+1} = \sum_{k=1}^{m} e_k \tag{10-18c}$$

$$w_i \geq 0 \qquad i = 1, \ldots, n+m+u+1 \tag{10-18d}$$

where the w_i, $(i = n+1, \ldots, n+m+u+1)$ are non-negative slack variables added to change consideration of an inequality constraint to consideration of an equality constraint in the optimization. [The reason why equalities in Equation (10-15b) were first eliminated and then restored is so that a slack variable is introduced for every equation. These slack variables enable an initial solution (set of values for w_i, $i = 1, \ldots, n+m+u+1$), from which to begin the search for the optimum solution, as will be seen shortly.] There are now $m+u+1$ equations in $n+m+u+1$ unknowns in Equations (10-18b), (10-18c), and (10-18d), referred to as the *constraint* set of equations. The Simplex method must begin its search for the optimum [maximum of Equation (10-18a)] from an initial solution that is not obvious. An initial "feasible" solution (one that satisfies the constraint set) may be obtained from an optimization similar to Equations (10-18):

max $-v$ subject to (10-19a)

$$\sum_{i=1}^{n} \alpha_{k,i} w_i \ + \ w_{n+k} \qquad\qquad = \ e_k \qquad k = 1, \ldots, \ m+u \qquad (10\text{-}19b)$$

$$\sum_{k=1}^{m}\sum_{i=1}^{n} \alpha_{k,i} w_i \ - \ w_{n+m+u+1} \ + \ v \ = \ \sum_{k=1}^{m} e_k \qquad\qquad (10\text{-}19c)$$

$$w_i \ \geq \ 0 \qquad i = 1, \ldots, \ n+m+u+1 \quad (10\text{-}19d)$$

$$v \ \geq \ 0 \qquad\qquad\qquad\qquad (10\text{-}19e)$$

where v is an "artificial" variable introduced as a computational device. The maximization of $-v$ corresponds to the minimization of v. If the minimum occurs at $v = 0$, then the solution to Equations (10-19) is feasible in Equations (10-18), and the search can begin in Equations (10-18) from this solution.

The Simplex method first solves for the $m + u + 1$ slack variables in Equations (10-19) as functions of the n original variables and artificial variable with the Gauss-Jordan method of elimination (see Chapters 6 and 7) and sets these $n + 1$ variables equal to zero. The $m + u + 1$ variables are referred to as *basic* variables and the other $n + 1$ (zeroed) variables are referred to as *nonbasic* variables. This will always give a feasible solution to Equations (10-19) if it exists, but it may not be an optimum solution [i.e., it will not necessarily be a maximum of Equation (10-19a)]. The objective function must always be written in terms of the nonbasic variables so that simple inspection of coefficients can be used in the Simplex method (following). Therefore, by solving Equation (10-19c) for the artificial variable and substituting it into Equation (10-19a), Equations (10-19) become

$$\max \left[\sum_{k=1}^{m}\sum_{i=1}^{n} \alpha_{k,i} w_i \ - \ w_{n+m+u+1} \ - \ \sum_{k=1}^{m} e_k \right] \text{ subject to} \qquad (10\text{-}20a)$$

$$\sum_{i=1}^{n} \alpha_{k,i} w_i \ + \ w_{n+k} \qquad\qquad = \ e_k \qquad k = 1, \ldots, \ m+u \qquad (10\text{-}20b)$$

$$\sum_{k=1}^{m}\sum_{i=1}^{n} \alpha_{k,i} w_i \ - \ w_{n+m+u+1} \ + \ v \ = \ \sum_{k=1}^{m} e_k \qquad\qquad (10\text{-}20c)$$

$$w_i \ \geq \ 0 \qquad i = 1, \ldots, \ n+m+u+1 \quad (10\text{-}20d)$$

$$v \ \geq \ 0 \qquad\qquad\qquad\qquad (10\text{-}20e)$$

The Simplex method selects the next feasible solution by allowing one of the $n + 1$ nonbasic variables to become positive (enter the *basis*, or set of basic variables) while letting one of the basic variables go to zero (leave the basis). The variable entering the basis is chosen as that variable which increases the value of the objective function at the highest rate, or has the highest coefficient in Equation (10-20a). The variable leaving the basis is chosen as that variable most limiting the growth of the objective function as the variable goes to zero. It is determined by inspection of the appropriate constraint equation containing that variable. After both the entering and leaving variables are chosen, the equation set is rewritten so that the objective function is again a function of the nonbasic variables and so that each basic variable appears in one and only one constraint equation. This is accomplished through the Gauss-Jordan method of elimination. The next feasible solution is then selected in the same manner. Iterations stop in the Simplex method when none of the nonbasic variables would increase the objective function if allowed to become positive; the last solution is then the optimum.

When the optimization is complete for Equations (10-19), the solution is a feasible solution to the problem of Equations (10-18) if the value of the artificial variable v is zero. If v is not zero, or no feasible solution to Equations (10-19) exists, then there is no feasible solution to the problem of Equations (10-18). This means that the constraint set must be changed by eliminating the lowest-priority equation or inequality in Equations (10-15) and its corresponding member in Equations (10-18) and (10-19). This cycle of optimization of Equations (10-19) and elimination of the lowest-priority equation is repeated until a feasible solution is found to the problem of Equations (10-18), i.e., until the artificial variable in Equations (10-19) equals zero. At this point, the objective function in Equation (10-18a) is rewritten in terms of the nonbasic variables, and the Simplex method is applied in Equations (10-18) to the feasible solution obtained from the problem of Equations (10-19). However, the artificial variable is never allowed to enter the basis (become non-zero) in the subsequent iterations. Figure 10-1 depicts the optimization algorithm for this linear programming solution of Equations (10-15).

EXAMPLE

Consider again the example of Figure 9-4 for a forecast made on September 15, 1998. The 41 years from 1954 through 1994 from the historical meteorology records of Lakes Superior, Michigan, Huron, and Erie and Georgian Bay are again used to define the meteorology time series segments of the operational hydrology approach. This time, the objective of Equation (10-5), representing maximization of the probability of SON precipitation in the upper third of its range over the entire area (basins of Superior, Michigan, Huron, Georgian Bay, and Erie), is used instead of minimization of the sum of squared differences between the weights and unity. The 20 equations in Figure 9-4 are used with inspections of the forty-one 12.5-month times series in the historical record and from quantile estimates to construct the equations in Figure 9-5, repeated in Figure 10-2. (See Exercise A2-10 in Appendix 2.) Now, however, a line is added corresponding to the new objective of Equation (10-5), and transformed as in Equation (10-10). Thus, the first line in Figure 10-2 corresponds to the objective function of Equation (10-10). The reader can verify this by using Equation (10-10) and inspection of column 8 in Table 9-2 with the 2/3 quantile defined in column 8 in Table 9-4.

Solution of the equations in Figure 10-2, by means of the Simplex method of linear programming is presented in Table 10-1. (All computations are with probabilities, both reference quantiles and forecasts, significant to three digits after the decimal point.) Note in Table 10-1 that several meteorology time series segments are unused (there are zero weights), but all of the meteorology outlooks in Figure 10-2 are used, just as was the case before with the old objective function (minimization of sum of squared differences of weights with unity). By using this set of weights, the probabilistic lake level outlooks for Lakes Superior, Michigan-Huron (including Georgian Bay), and Erie are estimated in Table 10-2. See Exercise A2-10 in Appendix 2 for this example, worked with the software described there.

By comparing Table 10-2 with Table 9-6, one can see the effect of including the different objective in the solution of the equations of Figure 9-4. The differences are presented in Table 10-3 to make the assessment more direct. While the numbers are small (they vary between -7 and 12 cm), such differences are significant for the Great Lakes, where small changes in lake levels correspond to huge changes in flows to and from the

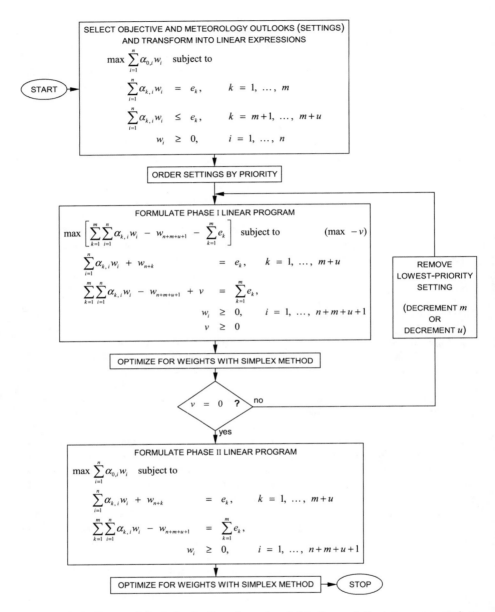

Figure 10-1. Determining physically relevant weights through linear programming.

lakes. Generally, the latter forecast is for higher lake levels, as might be expected with the objective of Equation (10-5); that is, the outlook of Table 10-2 appears higher than that of Table 9-6 more often than not and to a greater extent. The higher lake level outlooks appear mostly in the extended outlook (3 to 9 months into the future for Lake Superior and 6 to 12 months for Lakes Michigan-Huron and Erie) for the lower quantiles

$$\max \begin{bmatrix} 1000010000010000100000010000111110101100 \end{bmatrix} \times \begin{bmatrix} w_1 \\ w_2 \\ \vdots \\ w_{41} \end{bmatrix} \quad \text{subject to}$$

$$\begin{bmatrix} 11111111111111111111111111111111111111111 \\ 10100000110001010011001010010000000100001 \\ 01000101000100001100000100001101001011000 \\ 00000100000110000010101001100001100101110 \\ 10101011110000101101000100000110000000001 \\ 01100000111010000100111001000000000100000 \\ 10000101000100001000000010000111101011100 \\ 00010100000001010010101000110000101101110 \\ 10101011010000101101010010000100000010001 \\ 01100010111000000110111001100000000000000 \\ 10010100000100001000000011000010111101100 \\ 00000100000011000010101010001100001011101110 \\ 11101011010000101101010000001101000000001 \\ 00101011111000000111001000000000010100001 \\ 10010100000111011000000010000110000101110 \\ 00000100000111000010101010110000000100110 \\ 00000011010000111101010000000101000000001 \\ 00100011111000000100111000100000010001001 \\ 11000100000000001010000100011101101010110 \\ 00010000000011010010101010001100000011100110 \\ 10001011010000101101010000000101000010001 \end{bmatrix} \times \begin{bmatrix} w_1 \\ w_2 \\ \vdots \\ w_{41} \end{bmatrix} = \begin{bmatrix} 41.000 \\ 11.603 \\ 15.703 \\ 15.703 \\ 11.603 \\ 11.603 \\ 15.703 \\ 14.883 \\ 12.423 \\ 11.603 \\ 15.703 \\ 13.653 \\ 13.653 \\ 12.423 \\ 14.883 \\ 13.653 \\ 13.653 \\ 12.833 \\ 14.473 \\ 13.653 \\ 13.653 \end{bmatrix}$$

Figure 10-2. Alternative representation of Equations (6-3) and (10-10) and the equations in Figure 9-4.

Table 10-1. Outlook weights from solution of equations in Figure 10-2 for Superior, Michigan, Huron, Georgian Bay, and Erie application areas.

Index, i (1)	Weight, w_i (2)	Index, i (3)	Weight, w_i (4)	Index, i (5)	Weight, w_i (6)
1	0	15	0	29	1.230000
2	0	16	1.230000	30	0
3	0	17	0	31	3.339222
4	0	18	1.137067	32	1.467800
5	0	19	2.029955	33	1.230000
6	0	20	0	34	0.344400
7	1.216334	21	0	35	3.325555
8	0	22	5.686244	36	4.569222
9	0	23	0.210467	37	1.230000
10	0	24	2.994822	38	0
11	0	25	0.237800	39	0.554867
12	4.317755	26	0	40	0
13	0	27	1.322934	41	2.915555
14	0	28	0.410000		

Table 10-2. Upper Great Lakes probabilistic outlook using meteorology outlooks in Figure 10-2.

Month	Nonexceedance quantiles						
	10%	20%	30%	50%	70%	80%	90%
(1)	(2)	(3)	(4)	(5	(6)	(7)	(8)
Lake Superior water level (m above mean sea level)							
Sep 98	183.32	183.32	183.32	183.33	183.33	183.34	183.34
Oct 98	183.25	183.26	183.29	183.32	183.32	183.33	183.35
Nov 98	183.21	183.25	183.26	183.28	183.30	183.33	183.33
Dec 98	183.13	183.18	183.20	183.22	183.27	183.28	183.31
Jan 99	183.04	183.08	183.11	183.13	183.18	183.19	183.26
Feb 99	182.98	183.00	183.04	183.06	183.09	183.09	183.18
Mar 99	182.95	182.99	183.00	183.03	183.04	183.07	183.13
Apr 99	182.96	182.97	183.01	183.04	183.10	183.11	183.14
May 99	183.06	183.07	183.07	183.15	183.18	183.19	183.22
Jun 99	183.14	183.18	183.19	183.24	183.26	183.27	183.32
Jul 99	183.22	183.29	183.29	183.33	183.35	183.37	183.40
Aug 99	183.26	183.32	183.33	183.35	183.41	183.43	183.46
Sep 99	183.26	183.29	183.30	183.40	183.43	183.44	183.51
Lake Michigan-Huron-Georgian Bay water level (m above mean sea level)							
Sep 98	176.67	176.68	176.68	176.69	176.70	176.70	176.71
Oct 98	176.50	176.54	176.59	176.61	176.63	176.65	176.66
Nov 98	176.43	176.45	176.49	176.55	176.56	176.60	176.61
Dec 98	176.31	176.38	176.38	176.47	176.50	176.52	176.57
Jan 99	176.23	176.27	176.29	176.41	176.44	176.46	176.51
Feb 99	176.20	176.22	176.23	176.34	176.40	176.40	176.47
Mar 99	176.23	176.24	176.29	176.32	176.42	176.45	176.46
Apr 99	176.26	176.30	176.32	176.44	176.50	176.51	176.55
May 99	176.38	176.39	176.42	176.51	176.56	176.57	176.65
Jun 99	176.44	176.48	176.49	176.55	176.63	176.69	176.71
Jul 99	176.43	176.53	176.54	176.55	176.68	176.71	176.71
Aug 99	176.45	176.48	176.52	176.56	176.66	176.69	176.71
Sep 99	176.37	176.41	176.47	176.50	176.61	176.61	176.72
Lake Erie water level (m above mean sea level)							
Sep 98	174.45	174.46	174.46	174.46	174.47	174.47	174.48
Oct 98	174.26	174.27	174.27	174.29	174.31	174.32	174.37
Nov 98	174.15	174.15	174.18	174.21	174.26	174.30	174.33
Dec 98	174.10	174.11	174.12	174.18	174.31	174.31	174.33
Jan 99	174.09	174.12	174.12	174.22	174.33	174.34	174.36
Feb 99	174.17	174.19	174.19	174.21	174.29	174.30	174.31
Mar 99	174.17	174.23	174.28	174.29	174.36	174.42	174.46
Apr 99	174.28	174.29	174.31	174.39	174.47	174.49	174.53
May 99	174.33	174.38	174.39	174.43	174.50	174.52	174.55
Jun 99	174.32	174.38	174.40	174.48	174.51	174.53	174.54
Jul 99	174.27	174.36	174.37	174.45	174.47	174.49	174.50
Aug 99	174.22	174.29	174.30	174.37	174.40	174.42	174.42
Sep 99	174.04	174.18	174.20	174.26	174.32	174.32	174.35

Table 10-3. Objective function change on Upper Great Lakes levels probabilistic outlook.

Month	Nonexceedance quantiles								
	3%	10%	20%	30%	50%	70%	80%	90%	97%
(1)	(2)	(3)	(4)	(5)	(6)	(7)	(8)	(9)	(10)
Lake Superior water level change (cm)									
Sep 98	1	0	0	0	0	0	1	0	0
Oct 98	0	−1	−2	0	1	0	0	−1	−1
Nov 98	8	−1	2	1	1	1	0	−5	−1
Dec 98	10	0	2	2	2	4	0	−2	0
Jan 99	12	1	1	2	1	1	0	2	1
Feb 99	11	2	0	3	2	0	−1	2	2
Mar 99	7	5	5	5	4	1	0	1	0
Apr 99	3	4	3	5	3	2	1	1	0
May 99	4	3	1	1	7	2	0	0	−1
Jun 99	3	0	3	2	3	0	0	0	0
Jul 99	1	2	7	3	3	0	0	−1	0
Aug 99	1	1	4	2	0	1	0	−1	−2
Sep 99	1	−2	−3	−3	2	1	−1	−1	−1
Lake Michigan-Huron-Georgian Bay water level change (cm)									
Sep 98	0	0	0	0	0	1	0	0	0
Oct 98	−1	−4	−2	0	1	1	2	1	2
Nov 98	1	−2	−2	−1	2	−1	0	0	3
Dec 98	0	−4	0	−3	1	0	−1	0	2
Jan 99	3	0	−2	−2	3	1	0	2	1
Feb 99	2	1	−1	−3	3	0	−1	3	1
Mar 99	7	8	1	4	2	2	3	1	0
Apr 99	11	3	4	2	8	4	2	4	1
May 99	2	8	4	2	5	0	1	3	0
Jun 99	1	12	4	3	3	1	4	0	2
Jul 99	0	9	8	3	0	1	2	0	−1
Aug 99	0	12	3	2	1	0	1	1	0
Sep 99	0	8	3	0	−2	1	−1	0	1
Lake Erie water level change (cm)									
Sep 98	0	0	1	1	0	0	0	0	1
Oct 98	4	1	0	0	−1	−1	−2	0	1
Nov 98	1	2	−1	1	−1	−2	0	0	−1
Dec 98	1	1	−1	−6	−5	4	0	−3	0
Jan 99	−1	1	−1	−4	0	5	0	0	−1
Feb 99	−3	9	3	1	−1	−1	−2	−7	−2
Mar 99	4	1	5	5	1	3	2	4	2
Apr 99	4	7	1	0	2	4	3	1	2
May 99	0	7	5	0	0	2	4	0	−1
Jun 99	0	12	5	3	4	2	2	0	−2
Jul 99	7	10	7	1	4	0	−1	−1	−3
Aug 99	1	11	6	0	0	0	0	−2	−2
Sep '99	9	0	2	0	−1	−1	−1	−2	−1

(3% through 70% for Superior, 3% through 50% for Michigan-Huron, and 3% through 20% for Erie).

Even though one can exactly match meteorology outlooks in a derivative probabilistic outlook, that does not mean that the derivative outlook contains no other errors. Modeling and sampling errors still exist. In fact, sampling errors can be pronounced if many meteorology time series segments are weighted by zero and effectively eliminated from the sample. This happened in the outlook of Tables 10-1 and 10-2. In comparing the outlook with an earlier one in Table 9-6, which matched the same meteorology outlooks, the comparisons in Table 10-3 revealed inconsistent variations that undoubtedly result from the small effective sample that resulted. Only 21 of 41 segments were used (had nonzero weights in Table 10-1).

MULTIPLE SOLUTIONS

In the optimization of Equations (10-15), all expressions are *linear*, including the objective function [Equation (10-15a)]. This allows an alternative optimization technique (linear programming) to be used as compared with the earlier formulations of Equations (7-5) and (8-11), which used classical differential calculus solutions for zero slope of the Lagrangian. This formulation proves superior in its ability to include many alternative objectives (expressed as maximization of selected event probabilities).

Also, the formulation of Equations (10-15) allows non-negativity constraints on the weights to be *explicitly* included. This means that *all* solutions can be searched. The formulations of Equations (7-5) and (8-11), because they lack explicit inclusion of non-negativity constraints for the weights, look only at optimum solutions and discard them and lowest-priority constraints if nonnegativity constraints are unsatisfied. They also discard the many other possibilities that, while not optimum, might satisfy all constraints. Thus, the formulation of Equations (10-15), because it considers nonnegativity constraints explicitly, looks at *all* solutions within the framework of a single optimization without throwing away possibilities as before.

There is a trade-off, however. Multiple-optima solutions are now a possibility that did not exist before. Consider the following example, constructed as a variation of Equations (10-3)

$$\max \hat{P}\left[T_{\text{Sep}} \leq \hat{\tau}_{\text{Sep}, 0.333}\right] \text{ subject to} \qquad (10\text{-}21a)$$

$$\hat{P}\left[T_{\text{Sep}} \leq \hat{\tau}_{\text{Sep}, 0.333}\right] \leq 0.333 \qquad (10\text{-}21b)$$

$$\hat{P}\left[\hat{\tau}_{\text{Sep}, 0.333} < T_{\text{Sep}} \leq \hat{\tau}_{\text{Sep}, 0.667}\right] \leq 0.334 \qquad (10\text{-}21c)$$

$$\hat{P}\left[T_{\text{Sep}} > \hat{\tau}_{\text{Sep}, 0.667}\right] > 0.333 \qquad (10\text{-}21d)$$

Now, by inspection, the optimum is seen to occur along one of the constraints; i.e., the maximum of Equation (10-21a) occurs at the limiting value of the constraint in Equation (10-21b). Expressing in terms of Equations (10-15) where $\alpha_{0,i} = \alpha_{1,i}$, $i = 1, \ldots, n$,

$$\max \sum_{i=1}^{n} \alpha_{1,i} w_i \text{ subject to} \qquad (10\text{-}22a)$$

$$\sum_{i=1}^{n} \alpha_{1,i} w_i \leq 0.333\, n \qquad (10\text{-}22b)$$

$$\sum_{i=1}^{n} \alpha_{2,i} w_i \leq 0.334\, n \qquad\qquad (10\text{-}22\text{c})$$

$$\sum_{i=1}^{n} \alpha_{3,i} w_i \leq 0.333\, n \qquad\qquad (10\text{-}22\text{d})$$

$$w_i \geq 0 \qquad\qquad i = 1, \dots, n \qquad\qquad (10\text{-}22\text{e})$$

There are many solutions (values for w_i, $i = 1, \dots, n$) that are optimum. That is, any combination of weights that satisfy Equation (10-22b) as an equality rather than an inequality (an infinite set) maximizes Equation (10-22a). In this case, the solution would be expressed as all points (sets of weights) along the line segment (partial hyperplane) described by Equations (10-22b) and (10-22e). For other problems, there may be more than a single line segment defining the optima: a plane, a cube or block or other volume bounded by flat surfaces, or other higher-dimensioned shapes.

In the search algorithms employed in the linear programming solutions, these multiple optima can be detected (that is, the existence of more than a single optimum can be discerned) but the systematic exploration of them can be extensive. These multiple optima (while infinite in extent) can always be described as weighted combinations of a finite number of solution points. In practical terms, there are three limitations to the search for multiple optima: computation storage (or computer memory extent), computation time, and the growth of numerical error. The first two are easily appreciated. As multiple optimum solutions are discovered, they must be saved and compared with subsequent optimum solutions to avoid repeating their discovery. And of course, the searching and comparisons can take a great amount of time, depending upon the number of solution points. Numerical error is unavoidable on the computer and, while it can be minimized, can cause error growth. The linear programming algorithm involves a systematic search from one point in the constraint space to another, and an error in calculating one point can induce errors in the subsequent. While this appears manageable in most practical applications for finding the first optimum solution point in a problem such as Equations (10-15), it sometimes appears not to be manageable for finding a large number of additional optimum solution points when multiple optima are present. In fact, in many such problems, error growth can limit searches for other optima more than computer memory or time constraints.

Because the linear programming solution of Equations (10-15) may not be able to identify all multiple optima (solution points), the practitioner is faced with an interesting choice. He or she can either limit the number of optimum solution points that are returned by the optimization to the first few, or formulate the problem (choose an objective) so that the optimum is unique. In the first alternative, all optima are not generally found, so that some existing possibilities remain uninvestigated. It may be that some of these other possibilities would have been preferable (as determined from considerations outside of the problem) but remain unknown. In the second alternative, the practitioner may be facing the dilemma of electing not to solve the problem at hand, but instead (re)formulating the problem so that its solution behaves in a certain way. While this is a practical consideration, it is theoretically unappealing. Changing the problem so that it is solvable in a certain manner is not the same as solving the original problem. Nevertheless, there are many alternative, yet practical, problem formulations of interest, and the dilemma is not further addressed here. Note that, in the example of Figure 10-2 and Tables 10-1 and 10-2, the optimum is unique. If the objective function of Equation (10-6) is used instead of Equation (10-5) in defining the first line of Figure 10-2, there is again a unique optimum. If the objective function of Equation (10-7) is used instead of Equation (10-5) in defining the first line of Figure 10-2, there are multiple optima. See Exercise A2-11 in Appendix 2.

Chapter 11

EVALUATIONS

The methodology presented in Chapters 5–10 simply relates meteorology probability forecasts to hydrology models in an operational hydrology approach (by using the historical meteorology record) to generate hydrology probabilities in a meaningful way. There are some issues of performance that relate only to this methodology. Several choices can be made in applying the methodology (such as selection of meteorology forecasts, priority order, historical data, or solution objective) that impact on performance (such as the estimation of extremes or the goodness of forecast performance measures). While these performance issues can be discussed in general, exercises can only illuminate performance specific to an application and not general results for all applications. Generalized discussions in Chapters 5 through 10 address these issues. Readers will have to try variations of the methodology in their own applications to further evaluate the effects of these variations as these effects relate to them. They can easily accomplish this because the model simulations (to transform meteorology scenarios into hydrology scenarios) can be reused (not regenerated each time) for different sets of weights, as mentioned in the section Sampling and Modeling Independence in Chapter 6. Model simulations are often computationally expensive or time-consuming, while the calculation of weights is relatively easy (with the free software described in Appendix 2). The practitioner can test variations in the methodology (e.g., alternative meteorology forecasts, priorities, and objectives) by calculating the corresponding weights for each alternative in his or her application forecast and by applying them to the hydrology and meteorology scenarios.

Issues of performance other than those that relate to the methodology itself, as well as issues of methodology stability and error analysis, should also be investigated for practitioners to feel confident in applying the material contained in this book. This is a difficult point to address here in broad terms. These issues relate to the particular hydrology models and historical meteorology records that one actually uses in a given application. A practitioner can look at these issues and compare with other (outside) methods only in a specific application. It is not possible to generalize here for all variations of hydrology models or historical meteorology records that a practitioner might implement. However, the following examples illustrate some approaches to evaluating these issues, which may be useful as a guide in making other evaluations.

AN EXAMPLE DETERMINISTIC EVALUATION
The Coordinating Committee on Great Lakes Basic Hydraulic and Hydrologic Data commissioned the Great Lakes Environmental Research Laboratory (GLERL) in the autumn of 1997 to do an evaluation of present lake level forecasts and of GLERL's own Advanced Hydrologic Prediction System (AHPS). The objective was to *quickly* determine the *relative* suitability of the AHPS in making probabilistic outlooks of water levels. Available information consisted of actual and forecast Great Lake water levels from January 1995 (when the current meteorology probability forecasts became generally

available) through August 1997 (the time of the evaluation). The Detroit District Army Corps of Engineers makes the US lake-level forecasts, and Environment Canada makes the Canadian lake-level forecasts; both also issue a coordinated forecast.

GLERL simulated probabilistic lake level outlooks over the period January 1995 through August 1997 (which represent above normal lake levels) by (1) using supply sequences directly from the historical record, (2) using hydrology models with meteorology sequences directly from the historical record, and (3) using models on meteorology sequences and weighting for weather forecasts. It used weights obtained from the National Oceanic and Atmospheric Administration (NOAA) 1- and 3-month outlooks of precipitation and temperature only, with the objective of minimizing the sum of squares as in Equation (7-5a), which is argued in Chapter 10 to give outlooks close to climatology. (The objective formulations and optimization techniques in Chapter 10 were not available at that time for the evaluation; now, they would be a natural consideration in new evaluations.) An example forecast is presented in Figure 11-1, plotted from Table 10-2. GLERL integrated the probabilistic forecast distributions for average monthly lake levels to get the distribution mean for each month of the forecast to simplify comparisons and to allow direct comparison with other deterministic forecasts. It used all first-month mean forecasts of average monthly lake levels with actual average levels to calculate comparisons between the two: root mean square error (RMSE), Bias, correlation ($\hat{\rho}$), maximum error (MaxErr), and skill.

$$RMSE = \sqrt{\frac{1}{n}\sum_{i=1}^{n}(\ell_i - f_i)^2} \tag{11-1}$$

$$Bias = \frac{1}{n}\sum_{i=1}^{n}(\ell_i - f_i) \tag{11-2}$$

$$\hat{\rho} = \frac{\frac{1}{n}\sum_{i=1}^{n}(\ell_i - \overline{\ell})(f_i - \overline{f})}{s_\ell s_f} \tag{11-3}$$

$$MaxErr = \max|\ell_i - f_i| \tag{11-4}$$

$$skill = \frac{1}{n}\sum_{i=1}^{n}\frac{|\ell_i - \hat{\mu}_j|}{\hat{\sigma}_j}|\ell_i - f_i| \left/ \frac{1}{n}\sum_{i=1}^{n}\frac{|\ell_i - \hat{\mu}_j|}{\hat{\sigma}_j}|\ell_i - \hat{\mu}_j| \right. \tag{11-5}$$

$$\overline{\ell} = \frac{1}{n}\sum_{i=1}^{n}\ell_i \tag{11-6}$$

$$\overline{f} = \frac{1}{n}\sum_{i=1}^{n}f_i \tag{11-7}$$

$$s_\ell = \sqrt{\frac{1}{n}\sum_{i=1}^{n}(\ell_i - \overline{\ell})^2} \tag{11-8}$$

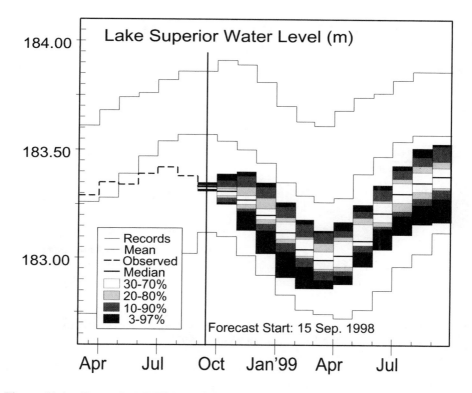

Figure 11-1. Example AHPS Lake Superior water level forecast for September 15, 1998.

$$s_f = \sqrt{\frac{1}{n} \sum_{i=1}^{n} \left(f_i - \bar{f} \right)^2} \qquad (11\text{-}9)$$

$$\hat{\mu}_j = \frac{1}{n_j} \sum_{i \in \Omega_j} \ell_i \qquad (11\text{-}10)$$

$$\hat{\sigma}_j = \sqrt{\frac{1}{n_j} \sum_{i \in \Omega_j} \left(\ell_i - \hat{\mu}_j \right)^2} \qquad (11\text{-}11)$$

where n = number of points of comparison of actual and forecast lake levels, ℓ_i = actual level i, f_i = forecast level i, j = month of the year corresponding to i, and Ω_j = all n_j jth months in the historical record. GLERL repeated this for months 2 through 6 as well.

RMSE, skill, and MaxErr are measures of the absolute differences between forecast and actual values. Skill is weighted to reflect the differences at extreme values more than differences near normal values. (Lower values of skill indicate better forecasts than higher values, as defined here.) Bias is a measure of the shift between the distributions of forecast and actual values. Correlation is a measure of how much variability in the

149

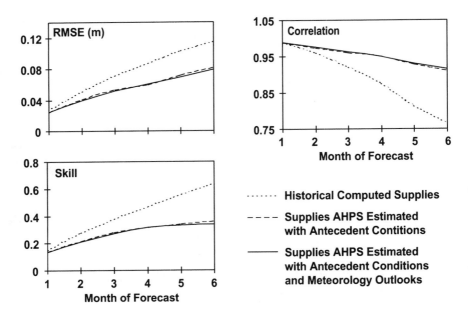

Figure 11-2. Lake Superior water level AHPS forecast improvements for January 1995 through August 1997.

actual values is explained by the forecast values. That is, it is a measure of how well the timing of variability is captured by the forecast method.

GLERL compared each of its forecast versions with the others to find the effects of considering antecedent moisture and heat storage conditions, and of considering weather forecasts; typical results are shown in Figure 11-2. The results suggest that, in this example, considering antecedent conditions greatly improved forecasts except in a few cases not shown here. Considering probabilistic weather information, while less dramatic in this example, also improved forecasts in the estimation of extremes but had less effect for near normal outlooks.

GLERL also compared the AHPS with the existing (operational) forecasts (United States, Canadian, & coordinated). Examples are pictured in Figure 11-3. GLERL's AHPS generally had lower *RMSE*, higher correlation, better skill (lower values), and lower maximum error than the United States, Canadian, and coordinated forecasts of lake levels. This suggests that GLERL's AHPS generally had the smallest differences with actual levels each month of the forecast, best captured the timing of variations of lake levels, and was most-consistently best at the extremes over different periods. However, GLERL's AHPS often appeared more biased than at least one of the other three methods. This suggests that the GLERL AHPS at the time of the evaluation was generally underpredicting slightly during that time of high levels, but the other forecasts were less consistent from period to period.

AN EXAMPLE PROBABILISTIC EVALUATION

Although the period of comparison is very short for comparison of probabilistic forecasts (only 32 months of forecast), GLERL did compare the various forecasts probabilistically. Shown in Figure 11-4 on the abscissa are the forecast non-exceedance prob-

150

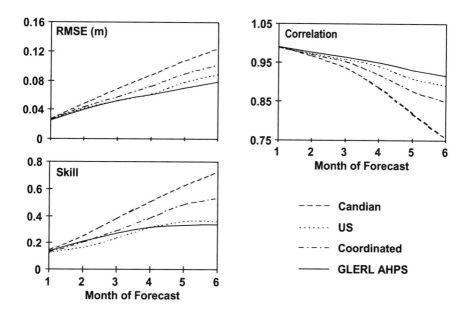

Figure 11-3. Lake Superior water level alternative forecast comparisons for January 1995 through August 1997.

abilities or cumulative distribution function (CDF); see Equation (2-9). The forecast lake level associated with each probability was compared to the actual lake level for each forecast; the fraction of actual levels that were at or below the forecast level are plotted on the ordinate. The dashed line represents a perfect probabilistic forecast. Also shown are the coordinated 5% and 95% non-exceedance levels.

On Lake Superior, the GLERL forecast did better probabilistically for all lags except lag 2. Remember, though, that this is a small sample for comparison. Note, furthermore, that both methods consistently underpredicted levels at longer lags, reflecting the increasing bias already observed (for the 6th month, for example, the median forecast is not exceeded only about 30% of the time, or exceeded about 70% of the time).

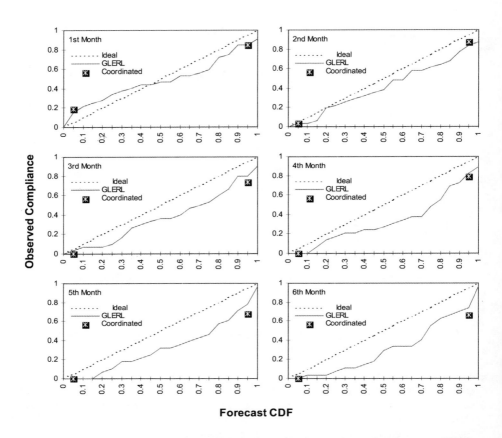

Figure 11-4. Lake Superior probabilistic forecast comparisons for January 1995 through August 1997.

Appendix 1

NOTATION

A	event label, as in "event A consists of all outcomes ω such that $X(\omega) \leq x$"
a	a real number constant, as the degree-day snowmelt constant in Equations (5-2) and the probability limit in Equations (8-21)
a_g	*Climate Outlook* probability setting for average air temperature for period g in the lower third of its 1961–90 range; see Equation (4-2a)
B	event label, as in "event A is contained in event B"
BIAS_{XZ}	bias, the expected value of the difference between the random variables X and Z, $(X - Z)$, used to measure the difference between the centers of their distributions; see Equation (2-17c)
$\widehat{\text{BIAS}}_{XZ}$	sample bias, the sample average of the difference between the random variable values x and z, $(x - z)$, used as an estimate of BIAS_{XZ}; see Equation (2-20c)
b_g	*Climate Outlook* probability setting for average air temperature for period g in the upper third of its 1961–90 range; see Equation (4-2b)
c_g	*Climate Outlook* probability setting for average precipitation for period g in the lower third of its 1961–90 range; see Equation (4-2d)
D_i	snow water equivalent at the end of day i
d	the derivative operator; see Equation (2-13a) for an example
d_g	*Climate Outlook* probability setting for average precipitation for period g in the upper third of its 1961–90 range; see Equation (4-2e)
$E[\cdot]$	the "expected value" operator, defined as the integral of the operand over the probability distribution function; see Equations (2-15)
e_k	selected weights sum limit in kth equation in Equations (6-7), (6-10), and (6-11), corresponding to a climate outlook probability setting, as, for example, in Equations (6-8)
$F(x)$	cumulative distribution function, the probability that $X \leq x$, as in Equation (2-19)
$f_X(x)$	the probability density function, the derivative of the cumulative distribution function, as in Equation (2-13a)
$f(x)$	penalty function, for use in optimization objective functions of Equations (8-22) and (8-23), for negative or zero arguments; see Equations (8-24)
f_i	forecast average lake level for month i; see Equations (11-1)–(11-11)

153

$g(x)$	penalty function, for use in optimization objective functions of Equations (8-22) and (8-23), for negative arguments; see Equations (8-24)
L	objective function (the Lagrangian) for an unconstrained optimization reformulated from the objective function for a constrained optimization by incorporating the constraints; see Equations (7-6) and (8-16)
ℓ_i	actual average lake level for month i; see Equations (11-1)–(11-11)
M_g	average basin moisture over period g
MSE_{XZ}	mean square error, the expected value of the square of the difference between the random variables X and Z, $(X - Z)^2$, used to characterize the difference between their values; see Equation (2-17d)
$\widehat{MSE}_{XZ}$	sample mean square error, the sample average of the square of the difference between the random variable values x and z, $(x - z)^2$, used as an estimate of MSE_{XZ}; see Equation (2-20d)
m	an integer constant, as the order of a value in a random sample in Equation (2-24) or (5-13), or the number of equalities to be satisfied in outlooks derivative to meteorology outlooks in Equations (6-10), (6-11), and (8-10)
m_g	a value of average basin moisture for period g; see Equation (3-13) or (3-15)
$\hat{m}_{g,\gamma}$	reference average basin moisture γ-probability quantile for period g; see Equations (3-12) and (3-14)
N	number of duplicated scenarios in the hypothetical very large structured set used for estimation in the operational hydrology outlook
N_L	number of duplicated scenarios, in the hypothetical very large structured set used for estimation in the operational hydrology outlook, which have $Q_{Sep} \leq \hat{\theta}_{Sep, 0.333}$ in Equation (5-21a)
n	number of scenarios available for use in generating the operational hydrology outlook; see Equation (5-8)
$n_{(A)}$	number of scenarios, available for use in generating the operational hydrology outlook, in which event A occurs (A is true) in Equations (5-27) and (5-28)
n_L	number of scenarios, available for use in generating the operational hydrology outlook, which have $Q_{Sep} \leq \hat{\theta}_{Sep, 0.333}$ in Equation (5-22a)
$n_{X \leq x}$	number of observations that are less than or equal to x, used in estimating the probability of nonexceedance with the sample relative frequency; see Equation (2-22)
$P[\cdot]$	probability of the event in brackets, representing its likelihood
$P[X \leq x]$	cumulative distribution function, the probability that $X \leq x$, as in Equation (2-9)
$\hat{P}[\cdot]$	relative frequency in a set, of the event in brackets, used as a probability estimate

p	an integer constant, as the number of strictly-less-than probability constraint inequalities selected from probabilistic meteorology outlooks for use in making an operational hydrology outlook; see Equations (8-9) and (8-10)
Q_g	total precipitation over period g
$Q_{h,g}$	total precipitation in application area h over period g, as $Q_{\text{Sup, SON98}}$ represents the Lake Superior application area and the September-October-November 1998 period
q	an integer constant, as the number of less-than-or-equal-to probability constraint inequalities selected from probabilistic meteorology outlooks for use in making an operational hydrology outlook; see Equations (8-9) and (8-10)
$\left(q_g\right)_i$	value of Q_g in scenario i
R_g	total basin runoff over period g
r	number of interval limits defining the $r + 1$ intervals of the real line for a variable's values; used additionally to denote the hydrology model in Equation (6-24)
$\left(r_g\right)_i$	value of R_g in scenario i
$\mathbf{S}$	vector of modeled hydrological output variables for the Bayesian Forecasting System
$\mathbf{s}$	vector of values of modeled hydrological outputs for the Bayesian Forecasting System; see Equation (6-24)
s_X^2	sample variance, the sample average of the square of the difference between random variable values x and the sample mean $\bar{x}$, $\left(x - \bar{x}\right)^2$, used as an estimate of σ_X^2; see Equation (2-20b)
s_Z^2	sample variance, the sample average of the square of the difference between random variable values z and the sample mean $\bar{z}$, $\left(z - \bar{z}\right)^2$, used as an estimate of σ_Z^2; see Equation (2-20b)
T_g	average air temperature over period g
$T_{h,g}$	average air temperature in application area h over period g, as $T_{\text{Sup, SON98}}$ represents the "Superior" application area and the September-October-November 1998 period
$\left(t_g\right)_i$	value of T_g in scenario i
u	an integer constant, as the number of probability constraint inequalities selected from probabilistic meteorology outlooks for use in making an operational hydrology outlook; see Equation (10-15c)
$\mathbf{u}$	vector of deterministic inputs to a hydrology model for the Bayesian Forecasting System
$\mathbf{V}$	vector of meteorology parameters of a probabilistic quantitative forecast for the Bayesian Forecasting System
$\mathbf{v}$	vector of values of meteorology parameters of a probabilistic quantitative forecast used in the Bayesian Forecasting System
v	"artificial" variable to drive to zero in Equations (10-19)
$\mathbf{W}$	vector of meteorology input variables for the Bayesian Forecasting System

W_g	average water surface temperature over period g
$\mathbf{w}$	vector of values of meteorology inputs for the Bayesian Forecasting System
$\left(w_g\right)_i$	value of W_g in scenario i
w_i	weight applied to ith scenario in the original set of possible future scenarios for calculation of estimates in a derivative outlook, or added "slack variable" to convert an inequality to an equation
X	a meteorology or hydrology variable
x_i	value for variable X in ith scenario in the set of possible future scenarios
x_k^N	value for variable X in kth duplicated scenario in the hypothetical very large structured set of N scenarios
x_i^n	value for variable X in ith scenario in the original set of n possible future scenarios
$\overline{x}$	sample mean, the sample average of random variable values x, used as an estimate of μ_X; see Equation (2-20a)
y_j	jth ordered value, from smallest to largest, for variable X in the set of possible future scenarios; see Equation (2-24)
y_j^N	jth ordered value, from smallest to largest, for variable X in the hypothetical very large structured set of N scenarios
y_j^n	jth ordered value, from smallest to largest, for variable X in the set of n possible future scenarios
Z	a meteorology or hydrology variable
z_j	value for variable Z in jth scenario in the set of possible future scenarios; used additionally in Equations (8-1) through (8-7) as the jth interval limit ($j = 1, \ldots, r$) defining the $r + 1$ intervals of the real line for a variable's values
z_i^n	value for variable Z in ith scenario in the original set of n possible future scenarios
$\overline{z}$	sample mean, the sample average of random variable values z, used as an estimate of μ_Z; see Equation (2-20a)
α_i	an integer coefficient equal to unity (1) if and only if the defining event is true in scenario i and zero (0) otherwise; see, for example the transforming of Equations (6-2) and (6-3) into Equations (6-4) and (6-5)
$\alpha_{k,i}$	an integer coefficient equal to unity (1) if and only if defining event k is true in scenario i and zero (0) otherwise; see, for example the transforming of Equations (6-6) and (6-3) into Equations (6-7) and (6-9); becomes then the coefficient in the kth equation on the ith weight (for the ith scenario) in Equation (6-7)
∂	the partial derivative operator; see Equations (7-7) and (8-17) for examples
γ	a real number (probability) between zero (0) and unity (1), inclusively
η	probability density of $\mathbf{W}$, or of $\mathbf{W}$ given $\mathbf{V}$, used in the Bayesian Forecasting System

ϑ_i duplication count for ith scenario in the original set of possible future scenarios for the hypothetical very large structured set

$\hat{\theta}_{g,\gamma}$ reference total precipitation γ-probability quantile for period g; see Equation (3-9)

$\hat{\theta}_{h,g,\gamma}$ reference total precipitation γ-probability quantile in application area h over period g, as $\hat{\theta}_{\text{Sup, SON98, 0.667}}$ represents the 0.667-probability quantile for the Lake Superior application area and the September-October-November 1998 period

λ_k LaGrange multiplier, representing the penalty associated with violation of the kth constraint equation in an optimization

μ_X mean, the expected value of random variable X used to define the central location of its distribution; see Equation (2-17a)

μ_Z mean, the expected value of random variable Z used to define the central location of its distribution; see Equation (2-17a)

ξ_γ reference γ-probability quantile for variable X

$\prod$ the product operator (product of expression following the operator indexed as indicated in the operator subscripts and superscripts); see Equation (2-19) for example

π probability density of $\mathbf{S}$ given $\mathbf{u}$ used in the Bayesian Forecasting System

ρ_{XZ} correlation, the expected value of the normalized product of the difference between random variable X and its mean $(X - \mu_X)$ and the difference between random variable Z and its mean $(Z - \mu_Z)$, used to characterize the linear dependence of one on another; see Equation (2-17e)

$\hat{\rho}_{XZ}$ sample correlation, the sample average of the normalized product of the difference between random variable values x and the sample mean $(x - \bar{x})$ and the difference between random variable values z and the sample mean $(z - \bar{z})$, used as an estimate of ρ_{XZ} ; see Equation (2-20e)

$\sum$ the summation operator (sum of expression following the operator indexed as indicated in the operator subscripts and superscripts); see Equation (2-14) for an example

σ_X^2 variance, the expected value of the square of the difference between random variable X and the mean $(X - \mu_X)^2$ used to define the spread of the distribution about its central location; see Equation (2-17b)

σ_Z^2 variance, the expected value of the square of the difference between random variable Z and the mean $(Z - \mu_Z)^2$ used to define the spread of the distribution about its central location; see Equation (2-17b)

$\hat{\tau}_{g,\gamma}$ reference average air temperature γ-probability quantile for period g; see Equation (3-5) for an example with $\gamma = 0.50$

$\hat{\tau}_{h,g,\gamma}$	reference average air temperature γ-probability quantile in application area h over period g, as $\hat{\tau}_{\text{Sup, SON98, 0.667}}$ represents the 0.667-probability quantile for the Lake Superior application area and the September-October-November 1998 period
ϕ	joint probability density of **W** and **V** for the Bayesian Forecasting System
ϕ_j	probability limit for a most-probable interval outlook for interval j
Ω	population or target space, the set of all possible outcomes in an experiment or the set of all possibilities in a population
ω	possibility or outcome, an element from Ω
$\forall$	"for all"; see Equation (2-2) for an example
$\int$	the integral operator (integral of the expression following the operator over the limits indicated in the operator subscripts and superscripts); see Equation (2-13b) for an example
$\varnothing$	the null set or empty set; see Equation (2-1d)
$\cap$	the intersection operator, denoting the intersection of sets; see Equation (2-1b) for an example
$\cup$	the union operator, denoting the union of sets; see Equation (2-1a) for an example
$\uplus$	the exclusive union operator, denoting the exclusive union of sets; see Equation (2-4a) for an example
^{c}	the complement operator, denoting the complement of a set; see Equation (2-1c) for an example
$\subset$	containment of one set in another; see Equation (2-2) for an example
$\in$	membership in a set; see Equations (2-1) for an example
$\Leftrightarrow$	"if and only if"; see Equation (2-2) for an example
$\Rightarrow$	"implies"; see Equation (2-2) for an example

Appendix 2

SOFTWARE DOCUMENTATION

All computations referred to in this book can be conveniently performed through an interactive graphical user interface, entitled *Derivative Outlook Weights*. This is a 32-bit *Windows* application designed for *Windows 95*, *Windows 98*, and *Windows NT* version 4.0. It is freely available over the World Wide Web from the Great Lakes Environmental Research Laboratory (GLERL); see the section Acquisition. This interface enables people to intuitively use the many probabilistic meteorology outlooks available in making their own derivative probabilistic outlooks. This appendix describes the software and its installation and use, provides helpful notes on generating outlooks with the software, summarizes the files in the interface package, and details the book's examples.

PURPOSE

The software provides the means to do three main tasks, essential in the construction of derivative outlooks from probabilistic meteorology outlooks. The first task is to enter all of the meteorology forecast information available to the decision maker, including all agency forecasts of meteorology probabilities and most-probable events, as well as probability forecasts that otherwise may be available. Part of this first task is also to sort through all of the entered information and arrange the forecast equations in a meaningful order of priority, enabling later computations to omit the lower-priority equations when necessary to effect a solution to the set of remaining equations. The second task is to define the parameters of the simulations used in the operational hydrology methodology to make the derivative forecast. These include the period of the forecast and the identification of the pieces of the historical meteorology record (up to 150 pieces allowed), which are to be used as possible scenarios in the model simulations. These two tasks can proceed independently, although it is relevant to use meteorology forecasts that are timely to the derivative forecast being made, and hence to the simulation parameters involved in the methodology. The third task is to construct the weights equations, representing the probabilistic meteorology outlooks, and to solve them for the weights to apply to the simulated outputs of the operational hydrology approach. The third task is dependent on the first two tasks.

These three tasks have been separated into three separate software applications, available as modules in the program documented in this appendix.

CONVENTIONS

Specific type styles in the graphical user interface *Derivative Outlook Weights* are used in this appendix to indicate the following items in this documentation, as illustrated:

Directories and filenames	(SUPERIOR, CLIMATE.SUP)
Program, module, and dialog box titles	**(Directory Selection)**
Menus and menu selections	(*Change Directory*)
Text in the program and modules main windows	("Current Directory")
Button commands and hints	("Simulations")

In this appendix, sidebars (boxed material) such as this denote special user instructions in a series of continuing exercises. Although they can be skipped in the casual reading of the material here, they are designed to quickly familiarize the user with the software interface. They are labeled for cross-reference elsewhere in this book as Exercises A2-1 through A2-4. In addition, a separate exercises section at the end of this appendix contains all of the worked examples used in this book.

ACQUISITION
The software may be acquired in a self-installing file (NOAAWGHT.EXE) by downloading from the following web site:

http://www.glerl.noaa.gov/wr/OutlookWeights.html

Inquiries may be made to

Thomas E. Croley II
Great Lakes Environmental Research Laboratory   313-741-2238
2205 Commonwealth Blvd.   FAX: 313-741-2055
Ann Arbor, MI 48105-2495   email: **Croley@glerl.noaa.gov**

INSTALLATION
The installation software installs *Derivative Outlook Weights* for *Windows 95*, *Windows 98*, or *Windows NT* version 4.0. It also installs example data (in subdirectory EXAMPLES DATA). If the exercises are run according to the instructions in the sidebars following, the results should agree with those sidebars. The following steps should ensure trouble-free installation:

1. After downloading or copying the file NOAAWGHT.EXE into a working directory of the user's choice, run NOAAWGHT.EXE to automatically install this product via an installation wizard. Choose the installation directory during the installation process. (The examples here presume that E:\OUTLOOKS is the installation directory. Substitute your installation directory in the examples that follow.)

2. *Optionally*, create another application area directory, if desired, to keep groups of user applications separate. The user will then be able to maintain different configurations of *Derivative Outlook Weights* for each separate group. This can be done at any later time or the default (EXAMPLE) can be used. (The program creates the subdirectory EXAMPLE the first time that the user extracts data from an archive in the subdirectory EXAMPLES DATA.) It may be convenient to create application area directories as subdirectories in the installation directory but it is not necessary. In any newly created application directory, place historical meteorology files, CLIMATE.???, for each application area desired in the group. This may be accomplished with the aid of appropriate commands within *Derivative Outlook Weights* (see the section Usage following.) The three-character extension (???) refers uniquely to an application area (e.g., in the examples archived in ZIP files contained in the subdirectory EXAMPLES DATA, CLIMATE.SUP refers to the Lake Superior basin application area, and CLIMATE.MIC refers to the Lake Michigan basin application area). See the section Files for more information on the CLIMATE.??? file.

3. *Optionally*, after successfully completing the installation, delete, if desired, all files in the installation directory except DERIVATIVE OUTLOOK WEIGHTS.EXE and UNZDLL.DLL. (However, the Windows uninstall procedure will be disabled.)

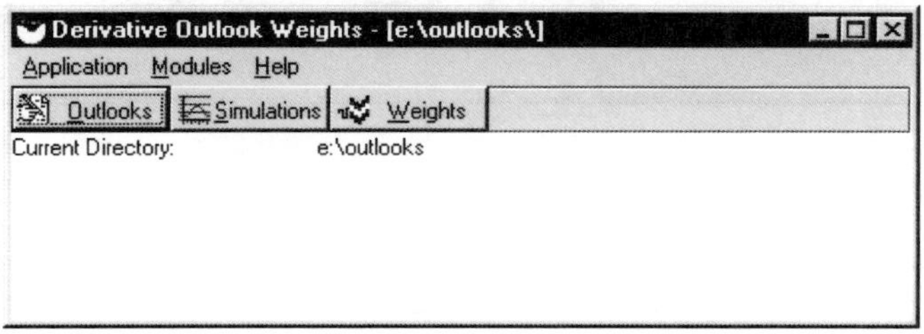

Figure A2-1. Derivative Outlook Weights initial window.

4. *Optionally*, to use old OTLKSETS.??? or OTLKEC3.??? files, created by earlier versions of this software, convert to the new formats by using appropriate commands from within *Derivative Outlook Weights* (see the section Usage). Note that ??? refers to a three-character extension uniquely denoting an application area. See the section Files for more information on the OTLKSETS.??? or OTLKEC3.??? files.

USAGE

The program *Derivative Outlook Weights* may be run by invoking it from the Windows "Start" menu, by double-clicking with the mouse on a desktop link icon, by double-clicking on the file name in Windows Explorer or Windows NT Explorer, by typing its name into the **Run** dialog box (accessible from the "Start" menu), or by typing its name at the command prompt in an MS-DOS console window. After invoking *Derivative Outlook Weights* the very first time, the user should be presented with a display similar to Figure A2-1. Figure A2-1 shows the default home directory as it appears if the software is installed into directory OUTLOOKS on the "E:" drive. This appears in the title of the window (as **Derivative Outlook Weights – [e:\outlooks\]**) and in the field marked "Current Directory" (as "e:\outlooks"). Figure A2-2 shows a typical view after the software has been used for a while. By selecting the *Application* drop-down menu on the main menu bar, as in Figure A2-3, and then selecting *Set Directory...* (denoted herein as *Application / Set Directory...*), the user invokes the **Directory Selection** dialog box (not shown). From there the user can select application directories in the standard Windows fashion. Simply double-clicking with the mouse anywhere in the main window of **Derivative Outlook Weights** also accesses this dialog box. Note that new directories or subdirectories can be created from within the *Derivative Outlooks Weights* program; they can also be deleted from within the program, but only if they are empty and not in use by other programs at the time. Delete by selecting the subdirectory in the **Directory Selection** dialog box and pressing the "Delete" key.

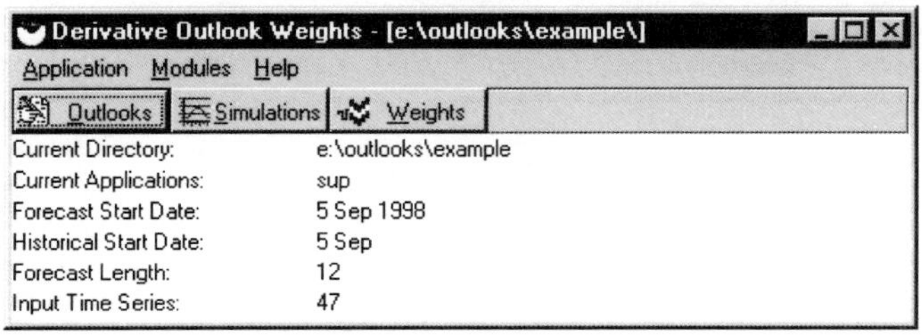

Figure A2-2. Derivative Outlook Weights example settings.

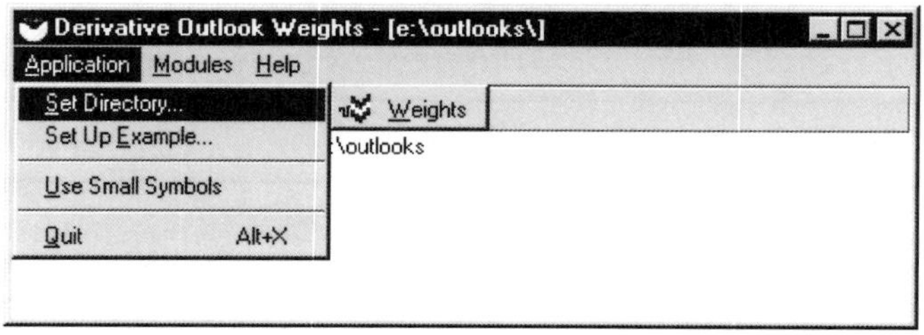

Figure A2-3. Derivative Outlook Weights Application menu.

Exercise A2-1

As an exercise, with the *Derivative Outlook Weights* program running, select *Application / Set Up Example...* to invoke the **Pick An Example** dialog box. Select "Example01" and click on the "accept selection" button on the left (let the cursor rest on any of the buttons for a hint to appear as to its function). This action should provide the **Successful Example Set Up** dialog box with a message indicating successful extraction of files. Click on the "OK" button. (Selecting "Example01" clears the EXAMPLE subdirectory and extracts AGNCSTRC, CLIMATE.SUP, and SIMULATE.$$$ into it from the EXAMPLE01.ZIP archive, located in the EXAMPLES DATA subdirectory. It also switches to the EXAMPLE subdirectory). The main window of **Derivative Outlook Weights** should be similar to Figure A2-2.

Climate Outlooks Settings Module

There are three modules that can now be invoked from the main window of **Derivative Outlook Weights**. They may be invoked from their respective buttons underneath the main menu bar, "Outlooks," "Simulations," and "Weights," or as submenu items under the *Modules* menu item on the main menu bar, *Outlooks*, *Simulations*, and *Weights*. Selecting these items invokes, respectively, the **Climate Outlooks Settings Module**, the **Simulation Settings Module**, and the **Mix Outlooks & Compute Weights Module**.

The first, the **Climate Outlooks Settings Module**, is invoked through the "Outlooks" button or by selecting *Modules / Outlooks*. An example display is given in Figure A2-4. The **Climate Outlooks Settings Module** also displays the application directory in its title. This module allows the user to set each application area for subse-

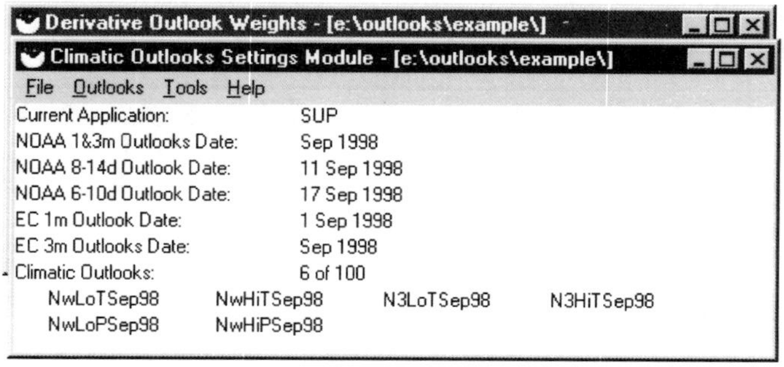

Figure A2-4. Climatic Outlooks Settings Module window.

162

quently entering available meteorological forecast information. It also allows outlook meteorology probabilities and agency outlook dates to be inputted, or previous inputs to be recalled, for five different agency forecasts as well as general probability statements (from additional agencies or user-defined). Finally, this module allows the user to arrange all of the outlook probability statements, for a selected application area, in a priority ordering for subsequent computations. Current values (if any) for the application area name, agency forecast dates, the number of outlooks used, and their priority order are displayed in the **Climate Outlooks Settings Module** window (see Figure A2-4). They appear in fields as follows. "Current Application" displays the application area name. "NOAA 1&3m Outlooks Date" displays the effective month and year of the National Oceanic and Atmospheric Administration (NOAA) Climate Prediction Center's (CPC's) *Climate Outlook*, one 1-month and thirteen 3-month forecasts of air temperature and precipitation probabilities. "NOAA 8-14d Outlook Date" displays the effective day, month, and year of the NOAA CPC's second-week, or 8–14 day, forecasts of air temperature and precipitation probabilities. "NOAA 6-10d Outlook Date" displays the effective day, month, and year of the NOAA CPC's day-6-to-day-10 (5-day) forecasts of the most-probable air temperature and precipitation events. "EC 1m Outlook Date" displays the effective day, month, and year of the Environment Canada (EC) Canadian Meteorological Centre's (CMC's) 1-month forecasts of the most-probable air temperature event. "EC 3m Outlook Date" displays the effective month and year of the first of EC CMC's four 3-month forecasts of the most-probable air temperature and precipitation events. And "Climatic Outlooks" displays the number of outlook probability statements to be used in making a derivative outlook and a listing of their priority order in an abbreviated format. These values may be set or reset only in the **Climate Outlooks Settings Module**.

Selecting An Application Area

By selecting *Outlooks / Select Area...*, shown in Figure A2-5, the user invokes the **Pick An Application** dialog box (not shown). From there the user can select, by check-marking, one application area in the standard Windows fashion. Also possible there, the user can alternatively define a new application area for which to enter forecast meteorology probabilities, by typing in a three-character mnemonic that refers uniquely to an application area. Simply double-clicking with the mouse in the "Current Application" field in the **Climate Outlooks Settings Module** window also accesses this dialog box.

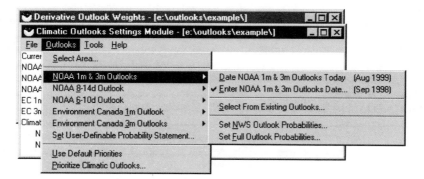

Figure A2-5. Climatic Outlooks Settings Module Outlooks menu.

Setting Effective Date of Agency Outlooks

This action is similar for each agency outlook, although the dates may be day, month, and year, or simply month and year, depending on the agency outlook being considered. By selecting *Outlooks / NOAA 8-14d Outlook*, *Outlooks / NOAA 6-10d Outlook*, *Outlooks / Environment Canada 1m Outlook*, or *Outlooks / Environment Canada 3m Outlooks*, a pop-up menu appears that is similar in function to that shown in Figure A2-5 for *Outlooks / NOAA 1m & 3m Outlooks*. Actions for all are similar to those described here for the latter.

By selecting either *Outlooks / NOAA 1m & 3m Outlooks / Date NOAA 1m & 3m Outlooks Today* or *Outlooks / NOAA 1m & 3m Outlooks / Enter NOAA 1m & 3m Outlooks Date...* , both of which are also shown in Figure A2-5, the user can set the effective date of the NOAA CMC *Climate Outlook* of one 1-month and thirteen 3-month forecasts of air temperature and precipitation probabilities to be used. Note that "effective date" is defined as the start date of the agency outlook (first) period, not to be confused with the issue date. The former selection sets the effective date to the month and year of the currently available *Climate Outlook* (depicted in Figure A2-5 as August 1999). The latter selection invokes the **Enter Month & Year** dialog box (not shown). This allows the user to set the effective date between January 1995 (when the first *Climate Outlook* was issued) and the forecast date possible on the current date (e.g., if the current date is July 21, 1999, then the August 1999 outlook is available). Simply double-clicking with the mouse in the "NOAA 1&3m Outlooks Date" field in the **Climate Outlooks Settings Module** window also accesses this dialog box. Likewise, double-clicking in any appropriate agency outlook date field in the main window accesses the date definition dialog boxes for that agency's outlook.

Selecting Existing Agency Outlooks

By selecting *Outlooks / NOAA 8-14d Outlook*, *Outlooks / NOAA 6-10d Outlook*, *Outlooks / Environment Canada 1m Outlook*, or *Outlooks / Environment Canada 3m Outlooks*, a pop-up menu appears that is similar in function to that shown in Figure A2-5 for *Outlooks / NOAA 1m & 3m Outlooks*. Actions for all are identical to that described here for the latter. By selecting *Select From Existing Outlooks...* from this pop-up menu, the user invokes the **Pick An Outlook** dialog box (not shown). Then the user can select from previously entered outlooks, if any exist.

Defining Agency Outlook Probabilities

This action is similar for NOAA's 1 and 3-month climate forecasts and NOAA's 8–14 day forecasts of event probabilities. By selecting *Outlooks / NOAA 8-14d Outlook*, a pop-up menu appears that is similar in function to that shown in Figure A2-5 for *Outlooks / NOAA 1m & 3m Outlooks*. Actions for both are similar to those described here for the latter. By selecting either *Outlooks / NOAA 1m & 3m Outlooks / Set NWS Outlook Probabilities...* or *Outlooks / NOAA 1m & 3m Outlooks / Set Full Outlook Probabilities...* , both of which are shown in Figure A2-5, the user invokes the **Select NOAA 1- & 3-Mo. Periods & Define Probabilities** dialog box (not shown). The former selection invokes it with all probabilities defined as incremental changes or "NWS probabilities," especially useful because the outlook maps are defined in terms of these probabilities; see Figures 4-1 through 4-4 or 4-6. (Simply double-clicking with the mouse in the "Climatic Outlooks" field in the **Climate Outlooks Settings Module** window brings up a pop-up menu, from which there is an entry that also accesses this dialog box.) The latter selection invokes the dialog box with all probabilities defined as absolute or "full probabilities;" see Equations (4-4) or Figure 4-5. This dialog box displays the one 1-month forecast and the thirteen 3-month forecasts, any of which can be defined by check-marking them. In both cases, when first check-marked, or when an existing check mark is first cleared and then re-check-marked, a dialog box is invoked to actually set the probabilities. In the first case (NWS probabilities), the dialog box is **Enter NWS Excess Likelihoods** (not shown); in the second case (full probabilities), the dialog box is **Enter 4 Probabilities** (not shown). In either case, the user can then enter relevant air temperature and precipitation probabilities from the NOAA *Climate Outlook* of one 1-month and thirteen 3-month forecasts or from the NOAA 8–14 day forecasts. Please note that as defined by NOAA, there are only four basic types of distributions allowed, as detailed in Chapter 4: (1) "above normal" (probability of high exceeds one-third with probability of low reduced accordingly), (2) "normal" (probability of middle range exceeds one-third with probabilities of being low and high reduced accordingly and equally), (3) "below normal" (probability of low exceeds one-third with probability of high reduced accordingly), and (4) "climatological" (probabilities of one-third in each range are used). When entering probability incremental changes ("NWS" probabilities), the dialog box automatically converts them correctly to one of these four distributions. When entering "full" probabilities, the dialog box uses them exactly as entered and it is users' responsibility to restrict themselves to these four distributions.

Defining Agency Outlook Most-Probable Events

This action is similar for NOAA's 6–10 day outlook, EC's 1-month outlook, and EC's extended 3-month outlooks of most-probable events. By selecting *Outlooks / NOAA 6-10d Outlook*, *Outlooks / Environment Canada 1m Outlook*, or *Outlooks / Environment Canada 3m Outlooks*, a pop-up menu appears that is similar in function to that shown in Figure A2-5 for *Outlooks / NOAA 1m & 3m Outlooks*, except that the last two items in the pop-up menu are replaced with a single item referring to the setting of the most-probable event. Actions for all are similar and are described here for the NOAA 6–10 day outlook. By selecting *Outlooks / NOAA 6-10d Outlook / Set NOAA 6-10d Most-Probable Event ...* (not shown), the user invokes the **Set NOAA 6-10d Most-Probable Event** dialog box (not shown). (Simply double-clicking with the mouse in the "Climatic Outlooks" field in the **Climate Outlooks Settings Module** window brings up a pop-up menu, from which there is an entry that also accesses this dialog box.) This dialog box allows the user to pick the temperature and precipitation intervals that are forecast as most probable.

User-Defined Outlook Probabilities

By selecting *Outlooks / Set User-Definable Probability Statement...* , which is also shown in Figure A2-5, the user invokes the **Define User Probability Statements** dialog box (not shown). This dialog box displays user-defined probability statements (initially the default is five), any of which can be defined by check-marking them. When first check-marked, or when an existing check mark is first cleared and then re-check-marked, the **Define User Probability Statement** dialog box (not shown) is invoked to actually set the probabilities. (Note that the former is plural and the latter is singular.) The user can then enter relevant variable type (temperature or precipitation), statement type (less than, less than or equal to, equal to, greater than or equal to, or greater than), reference quantile definitions, forecast dates, and probabilities. The beginning and ending dates may be any day from January 1, 1995, through December 31,

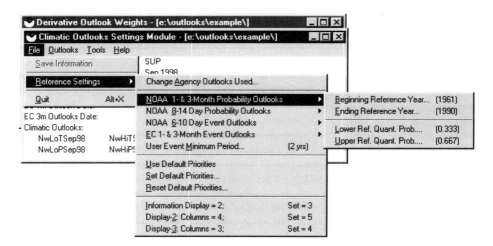

Figure A2-6. Climatic Outlooks Settings Module File menu.

of the year two years after the current year. (The two-year period is changeable by selecting *File / Reference Settings / User Event Minimum Period...* , shown in Figure
A2-6, which invokes the **Enter A Number** dialog box, not shown.) The user-defined
probability outlooks may correspond to other-agency outlooks, not directly supported in
the software, or to probability statements of the user's choosing. Simply double-clicking
with the mouse in the "Climatic Outlooks" field in the **Climate Outlooks Settings
Module** window brings up a pop-up menu, from which there is an entry that also accesses the **Define User Probability Statements** dialog box.

Defining Outlook Priorities
By checking *Outlooks / Use Default Priorities* (shown in Figure A2-5) or *File / Reference Settings / Use Default Priorities* (shown in Figure A2-6), the previously set
default ordering of probability statements is used to establish their priority order. These
default priorities may be defined (set) or reset (to values original to the software) by selecting *File / Reference Settings / Set Default Priorities...* or *File / Reference Settings / Reset Default Priorities...* , respectively (both are shown in Figure A2-6).
Both invoke the **Order by Priority (by dragging)** dialog box (not shown) for use in
defining the default priorities. By selecting *Outlooks / Prioritize Climatic Outlooks...* , shown in Figure A2-5, the user invokes the **Order by Priority (by dragging)** dialog box (not shown) for use in arbitrarily ordering the probability statements
instead of using default priorities. Then the user may simply drag the probability statements, displayed there, in standard Windows fashion to reorder them and to define
which are actually included in the ordering. Simply double-clicking with the mouse
below the "Climatic Outlooks" field in the **Climate Outlooks Settings Module** window also accesses the **Order by Priority (by dragging)** dialog box. Note that if the
user uses default priorities by selecting *File / Reference Settings / Use Default Priorities*, a check mark will appear next to that menu selection and the user will be unable
to arbitrarily order the probability statements in the **Order by Priority (by dragging)**
dialog box. If the user wishes to arbitrarily order probability statements, then he or she
must first reselect *File / Reference Settings / Use Default Priorities* to clear the
check mark there.

166

Saving Climatic Outlook Settings

Anytime an entry to one of the agency outlook definitions is made and accepted, it is stored on disk. The user may switch between application areas and enter forecast information for each without losing earlier entries. The "current" settings shown in the main window when exiting the module will be the settings recalled to the main window the next time the **Climatic Outlooks Settings Module** is invoked on the same application area. If any changes were made to the current settings shown in the main window since selection of the application area and not yet saved, the module will prompt the user to also save them separately for use by the third module (**Mix Outlooks & Compute Weights Module**) at exit time or when the user attempts to switch application areas. By selecting *File / Save Information*, when enabled, the user can save all current selections separately for use by the third module. If no selections were set or reset since the last time the **Climatic Outlooks Settings Module** was used on the present application area, then the *File / Save Information* menu item is disabled (grayed out) as shown in Figure A2-6. Whenever any of the fields are set or reset, the *File / Save Information* menu item is enabled.

Exercise A2-2

As a continuing exercise, run the *Derivative Outlook Weights* program or, if already running, exit from any module currently running, if necessary, to return to the **Derivative Outlook Weights** main window. Select *Application / Set Up Example...* to invoke the **Pick An Example** dialog box. Select "Example02," and click on the "accept selection" button on the left (let the cursor rest on any of the buttons for a hint to appear as to its function). (This action clears the EXAMPLE subdirectory; extracts files into it from the EXAMPLE02.ZIP archive, located in the EXAMPLES DATA subdirectory; and informs the user as to its successful completion.) The main window should appear similar to that shown in Figure A2-2. Then open the **Climate Outlooks Settings Module** by selecting either the "Outlooks" button or *Modules / Outlooks*. The window display should be blank. Select *Outlooks / Select Area...*, to invoke the **Pick An Application** dialog box, and select the "SUP" application, which is the only one shown. Click the "accept selection" button (leftmost button). The window should no longer be blank but should appear similar to Figure A2-4.

Exercise A2-2a

Select *Outlooks / NOAA 1m & 3m Outlooks / Date NOAA 1m & 3m Outlooks Today* to set the effective date to the latest available outlook. [The **Order by Priority (by dragging)** dialog box will appear. Simply click the "accept priority order" button (leftmost button).] Select *Outlooks / NOAA 1m & 3m Outlooks / Enter NOAA 1m & 3m Outlooks Date...* to invoke the **Enter Month & Year** dialog box, and enter September 1998. (Again, accept the priority order in the dialog box that appears.) Alternatively, select *Outlooks / NOAA 1m & 3m Outlooks / Select From Existing Outlooks...* to invoke the **Pick An Outlook** dialog box and select the September 1998 outlook already defined. (Again, accept the priority order in the dialog box that appears. Ordinarily, one would use only one of these three methods.) These actions are similar for other agency outlook definitions and will not be repeated in the following exercises (A2-2b through A2-2f). Note again that "effective date" is defined as the start date of the agency outlook (first) period, not to be confused with the issue date. Thus, the September 1998 outlook begins in September even though it was issued in mid-August.

Select _Outlooks_ / _NOAA 1m & 3m Outlooks_ / _Set NWS Outlook Probabilities..._ , to invoke the **Select NOAA 1- & 3-Mo. Periods & Define Probabilities** dialog box, and compare the settings there with Table 4-1. Clear and then re-check-mark the September setting to invoke the **Enter NWS Excess Likelihoods** dialog box and experiment with the settings there. When finished, exit by selecting the "cancel settings" button and then exit from the **Select NOAA 1- & 3-Mo. Periods & Define Probabilities** dialog box by selecting its "cancel" button. Next, select _Outlooks_ / _NOAA 1m & 3m Outlooks_ / _Set Full Outlook Probabilities..._ to invoke the **Select NOAA 1- & 3-Mo. Periods & Define Probabilities** dialog box (this time with "full" probabilities displayed) and compare the settings there with Table 4-1 and Figure 4-5. Clear and then re-check-mark the September setting to invoke the **Enter 4 Probabilities** dialog box and experiment with the settings there. When finished, exit by selecting the "cancel" button and then exit from the **Select NOAA 1- & 3-Mo. Periods & Define Probabilities** dialog box by selecting its "cancel" button. (Ordinarily, one would use only either _Set NWS Outlook Probabilities..._ or _Set Full Outlook Probabilities... ._) Repeat these steps to explore the other agency outlook definition dialog boxes. Always exit (for now) by selecting the "cancel" button; otherwise the **Order by Priority (by dragging)** dialog box is automatically invoked, as discussed in Exercise A2-2h.

Exercise A2-2b
Experiment, as in Exercise A2-2a, by selecting _Outlooks_ / _NOAA 8-14d Outlook_ / _Set NWS Outlook Probabilities..._ or _Outlooks_ / _NOAA 8-14d Outlook_ / _Set Full Outlook Probabilities..._ , and compare current settings for NOAA's 8–14 day outlooks of air temperature and precipitation probabilities with Equations (4-4) and Figure 4-6.

Exercise A2-2c
Experiment by selecting _Outlooks_ / _NOAA 6-10d Outlook_ / _Set NOAA 6-10d Most-Probable Event..._ to invoke the **Set NOAA 6-10d Most-Probable Event** dialog box and compare current settings already present for NOAA's 6–10 day outlooks of most-probable air temperature and precipitation with Equations (4-10) and Figure 4-10.

Exercise A2-2d
Experiment by selecting _Outlooks_ / _Environment Canada 1m Outlook_ / _Set EC 1m Most-Probable Event..._ to invoke the **Select EC 1-Month Most-Probable Event** dialog box and compare current settings for EC's 1-month outlooks of most-probable air temperature with Equations (4-12) and Figure 4-11.

Exercise A2-2e
Experiment by selecting _Outlooks_ / _Environment Canada 3m Outlook_ / _Set EC 3m Most-Probable Event..._ to invoke the **Select EC 3-Month Most-Probable Events** dialog box and compare current settings for EC's seasonal 3-month outlook of most-probable air temperature and precipitation with Equations (4-13) and Figure 4-12.

Exercise A2-2f
Experiment by selecting _Outlooks_ / _Environment Canada 3m Outlook_ / _Set EC 3m Most-Probable Event..._ to invoke the **Select EC 3-Month Most-Probable Events** dialog box and compare current settings for EC's extended seasonal 3-month outlooks of most-probable air temperature and precipitation on alternative tabbed pages with Figures 4-13 and 4-14.

Exercise A2-2g

Finally, select *Outlooks / Set User-Definable Probability Statement...* to invoke the **Define User Probability Statements** dialog box. Note that the equations shown there are Equations (4-9a) through (4-9e). Clear and re-check-mark the first equation to invoke the **Define User Probability Statement** dialog box (note that the latter dialog box title is singular while the former is plural). Experiment with defining various kinds of probability statements ("Variable Type" and "Statement Type"), the definition of the reference quantiles (lower quantile probability, upper quantile probability, period of the variable, and reference period), and the statement probability. Exit by selecting the "cancel" button from both of these dialog boxes.

Exercise A2-2h

Starting in the **Climate Outlooks Settings Module**, double-click anywhere in the window below the "Climatic Outlooks" field. This invokes the **Order by Priority (by dragging)** dialog box. If the screen will permit it, drag the lower right corner of the dialog box to enlarge it to the point that the scroll bars disappear. Look at the alternative display by selecting the second button from the left (again, let the cursor rest on any of the buttons for a hint to appear as to its function). Experiment by selecting groups of statements in the standard Windows fashion (e.g., click on one, then shift-click on another to select a range; add to them with a control-click on any others). Drag to reorder and define those which are used and unused. Before exiting, select the third button from the left to reset the order to that present upon entering this dialog box; then select the first button to exit. Finally, quit the **Climate Outlooks Settings Module** without saving anything, if prompted.

Converting Old Forecast Databases

There has been a change in format, over the last few years, in two of the software-constructed databases representing user-input values taken from agency forecast probability maps. They affect files OTLKSETS.??? and OTLKEC3X.???, used, respectively, for certain NOAA and EC outlooks (NOAA's 1- and 3-month climate forecasts of air temperature and precipitation event probabilities and EC's seasonal and extended 3-month outlooks of most-probable air temperature and precipitation events). Should the user have these agency forecasts defined under previous versions of this software, he or she can easily convert them for use with this software. Attempting to convert OTLKSETS.??? when unnecessary will not change anything, so the user may attempt it if unsure of the earlier version. If change is necessary, the software will create a new file of the same name and rename the old to OTLKSOLD.???. Attempting to convert OTLKEC3.??? to OTLKEC3X.??? (note the change in file name) will overwrite any existing OTLKEC3X.???, so the change should be made (if necessary) before any new information is entered with this software for the EC 3-month outlooks.

In the **Climate Outlooks Settings Module**, by selecting *Tools / Convert Old NOAA 1-3m Data...* the user invokes an **Instructions** dialog box explaining the conversion process for OTLKSETS.??? files. Should the user wish to proceed, click the "Yes" button; otherwise, click the "No" button. If the user proceeds and no OTLKSETS.??? files are found, a message will be given to the user to that effect and the conversion will terminate. If appropriate files are found, they will be converted if necessary and appropriate messages will be given. By selecting *Tools / Convert Old EC 3-m Data...* , the user will invoke an **Instructions** dialog box explaining the conversion process for OTLKEC3.??? files. The resulting actions are similar.

Setting Allowable Agency Outlooks

Although normally never accessed by most users, there are also many reference settings available in the **Climate Outlooks Settings Module**. Figure A2-6 shows menus to access them. *Be careful making changes to these reference settings! Do not change them unless you are sure you know what you are doing!* The impacts of reference setting changes can be far-reaching. By selecting *File / Reference Settings / Change Agency Outlooks Used...* , the user can invoke the **Change Agency Outlooks Used** dialog box (not shown) to configure which agency outlooks are to be used henceforth. Note that users always have access to the databases that are created for each set of agency outlooks (such as in Exercise A2-2), whether using them or not. That is, a user can enter information for any agency outlook through the menus shown in Figure A2-5, even if not planning to use that information subsequently (as defined in the **Change Agency Outlooks Used** dialog box). The default outlooks are all one 1-month and thirteen 3-month temperature and precipitation forecasts from NOAA (56 equations), all second-week (8–14 day) temperature and precipitation forecasts from NOAA (4 equations), all 6–10 day temperature and precipitation most-probable event forecasts from NOAA (8 equations), all 1-month temperature most-probable event forecasts from EC (3 equations), and all seasonal and extended 3-month temperature and precipitation most-probable event forecasts from EC (24 equations). These total 95 equations and 100 equations are allowed. The difference is made up of user-defined statements, and there are then five in the default configuration. While the user will have little occasion to change this configuration, the flexibility allows changes in the future as new agency outlooks become available and old ones are discontinued. Sometimes, a user may wish to temporarily expand the number of user-defined statements possible by eliminating agency outlooks, and then return to the default configuration (see Exercise A2-7). Finally, the user may reorder the statements within an agency grouping. This reordering affects the default priority ordering and is usually quite unnecessary.

Defining Agency Reference Settings

The reference years and reference quantile probability definitions for each agency, as defined in Chapter 4, are set in the default configuration of the software and may be changed as the agency changes them in the future. They are accessed through the menu shown in Figure A2-6. This action is similar for each agency outlook, although the quantile definitions differ from agency to agency. By selecting *File / Reference Settings / NOAA 8-14d Probability Outlooks*, *File / Reference Settings / NOAA 6-10d Event Outlooks*, or *File / Reference Settings / EC 1- & 3-Month Event Outlooks*, a pop-up menu appears that is similar in function to that shown in Figure A2-6 for *File / Reference Settings / NOAA 1- & 3-Month Probability Outlooks*. Actions for all are similar to those described here for the latter. By selecting *File / Reference Settings / NOAA 1- & 3-Month Probability Outlooks / Beginning Reference Year...* , the user invokes the **Enter Year** dialog box (not shown), where the beginning reference year can be set. The ending reference year can be set from a similar menu selection. Likewise, by selecting *File / Reference Settings / NOAA 1- & 3-Month Probability Outlooks / Lower Ref. Quant. Prob....* , the user invokes the **Enter Probability** dialog box (not shown), where the lower reference quantile probability can be set. The upper reference quantile probability can be set from a similar menu selection.

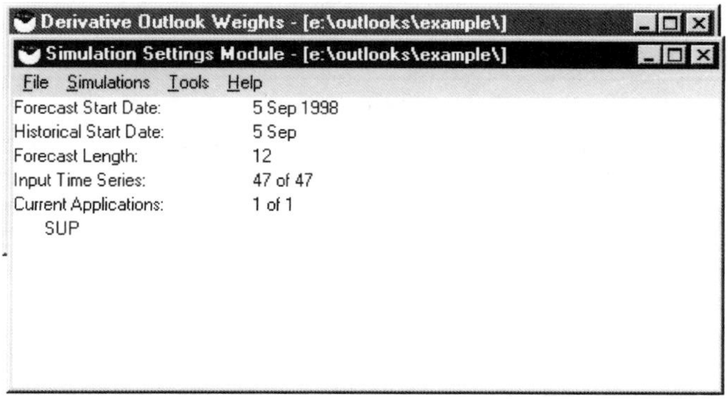

Figure A2-7. Simulation Settings Module window.

Simulation Settings Module

The second module callable from the main window of **Derivative Outlook Weights** is the **Simulation Settings Module** and is invoked through the "Simulations" button or by selecting *Modules / Simulations*. An example display is given in Figure A2-7. The **Simulation Settings Module** also displays the application directory in its title. This module allows the user to set the start date for the derivative forecast (the forecast that will be generated from model simulations and meteorological forecasts) and the start date for each meteorological scenario (beginning in each year of the historical record) used in the model simulations. It also allows input of the forecast length (length of the meteorology scenarios), which scenarios to use (which years of the historical record), and the application area(s) to be used. Current values, if any, are displayed for these settings in both the **Simulation Settings Module** window (see Figure A2-7) and the parent **Derivative Outlook Weights** main window (see Figure A2-2). In both, they appear respectively in the fields "Forecast Start Date," "Historical Start Date," "Forecast Length," "Input Time Series," and "Current Applications." These values may be set or reset only in the **Simulation Settings Module**.

Selecting Application Areas
By selecting *Simulations / Select Areas...* , as in Figure A2-8, the user invokes the **Pick Applications** dialog box (not shown). From there the user can select, by check-marking, one or more application areas (if respective CLIMATE.??? files are available in the current directory) in the standard Windows fashion. Simply double-clicking with the mouse in or below the "Current Applications" field in the **Simulation Settings Module** window also accesses this dialog box.

Setting Derivative Forecast Start Date
By selecting either *Simulations / Start Forecast At Data End* or *Simulations / Enter Forecast Start Date...* , both of which are also shown in Figure A2-8, the user can set the derivative forecast start date. The former selection sets the start date to the day after the earliest end of the data contained in the CLIMATE.??? files corresponding to the application area(s) selected. The latter selection invokes the **Enter Date** dialog box (not shown) and allows the user to set the start date to any day after the end of the data, but no later than the end of the year 2 years after the current date. Simply double-clicking with the mouse in the "Forecast Start Date" field in the **Simulation Settings Module**

171

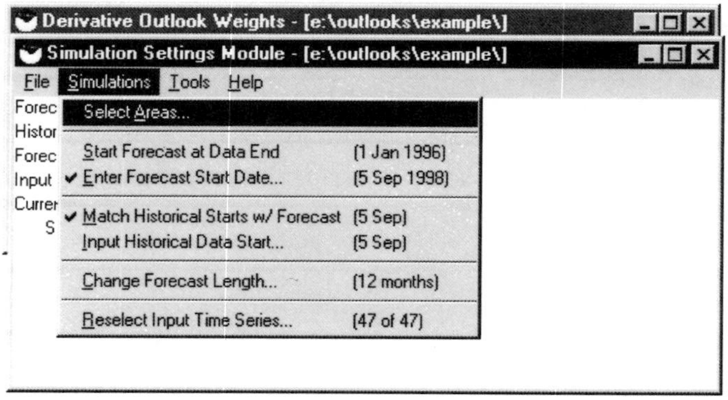

Figure A2-8. Simulation Settings Module Simulations menu.

window also accesses this dialog box. (The 2-year period is changeable by selecting *File / Reference Settings / Min. Forecast Start Period...* , which invokes the **Enter A Number** dialog box, not shown.)

Defining Historical Meteorology Time Series Sample

By selecting either *Simulations / Match Historical Starts w/ Forecast* or *Simulations / Input Historical Start Date...* , both of which are also shown in Figure A2-8, the user can set the historical start date. The former selection sets the historical start date for each meteorological scenario, beginning in each year of the historical record, to the day of the year corresponding to the selected forecast start date. The latter selection invokes the **Enter Abbreviated Date** dialog box (not shown) and allows the user to set the start date to any day of the year. Normally, this option is not used in practice. Simply double-clicking with the mouse in the "Historical Start Date" field in the **Simulation Settings Module** window also accesses this dialog box.

By selecting *Simulations / Change Forecast Length...* , which is also shown in Figure A2-8, the user invokes the **Enter A Number** dialog box (not shown). The user can then set the forecast length to any integral value between 1 month and the number of months possible in a meteorological scenario (determined from the data period common to all of the CLIMATE.??? files, corresponding to all the application areas chosen). Simply double-clicking with the mouse in the "Forecast Length" field in the **Simulation Settings Module** window also accesses this dialog box.

By selecting *Simulations / Reselect Input Time Series...* , which is also shown in Figure A2-8, the user invokes the **Pick Numbers** dialog box (not shown). The user then can check-mark which scenarios to use (which start years of the historical record to use). Available selections (start years) are taken from the data period common to all of the CLIMATE.??? files (only the first 150 selections are used). Simply double-clicking with the mouse in the "Input Time Series" field in the **Simulation Settings Module** window also accesses this dialog box.

Saving Simulation Settings

By selecting *File / Save Information*, when enabled, the user can save all selections made, to be used in the third module (**Mix Outlooks & Compute Weights Module**). If no selections were set or reset since the last time the **Simulation Settings Module** was used in the directory named in its title, then the *File / Save Information* menu item

172

is disabled (grayed out). Whenever any of the fields are set or reset, the *File / Save Information* menu item is enabled.

Exercise A2-3

As a continuing exercise, run the *Derivative Outlook Weights* program or, if it is already running, exit from any module currently running, if necessary, to return to the **Derivative Outlook Weights** main window. Select *Application / Set Up Example...* to invoke the **Pick An Example** dialog box. Select "Example03," and click on the "accept selection" button on the left (let the cursor rest on any of the buttons for a hint to appear as to its function). (This action clears the EXAMPLE subdirectory, extracts files into it from the EXAMPLE03.ZIP archive, located in the EXAMPLES DATA subdirectory, and informs the user of its successful completion). The main window should show the EXAMPLE subdirectory appearing in the "Current Directory" field (e.g., as "e:\outlooks\example") but be otherwise empty. Then open the **Simulation Settings Module** by selecting either the "*S*imulations" button or *Modules / Simulations*. The window display should be blank. Select *Simulations / Select Areas...* , to invoke the **Pick Applications** dialog box, and select the "SUP" application, which is the only one shown. Click on the "accept selection" button. The window should no longer be blank but will contain default values for all fields dependent on the contents of CLIMATE.SUP. Select *Simulations / Enter Forecast Start Date...* or double-click with the mouse on "Forecast Start Date" and enter the date September 5, 1998, and click the "OK" button. Ensure that the menu item, *Simulations / Match Historical Starts w/ Forecast*, is check-marked. Select *Simulations / Change Forecast Length...* or double-click with the mouse on "Forecast Length," enter 12 months if necessary, and click the "OK" button. Select *Simulations / Reselect Input Time Series...* or double-click with the mouse on "Input Time Series," check-mark all time series if not already check-marked, and click on the "OK" button. The display should correspond to Figure A2-7. Select *File / Quit* or otherwise try to close the **Simulation Settings Module** window and get a message asking to quit without saving the simulate information to a file. Select the "No" button to prevent this early exit. Select *File / Save Information* and then *File / Quit*. The display should now correspond to Figure A2-2.

User-Supplied Historical Climate Data

The user must supply a file of historical daily air temperatures and precipitation for each application area he or she defines. The file is CLIMATE.??? (where ??? denotes the user-defined three-character mnemonic for the application area); its structure and content are defined in the Files section of this appendix. To facilitate the building of this file, two tools are presented in the **Simulation Settings Module**. The first tool allows the user to supply an ASCII text file of historical daily data for air temperature and precipitation and to convert it easily and quickly to the required form. By selecting *Tools / Build CLIMATE.??? File...* , the user invokes the **Pick Applications** dialog box. From there the user can select, in standard Windows fashion, any of the applications represented in the current directory with ASCIDATA.??? files. These files are plain ASCII text files that can be created with simple editors or re-derived from existing CLIMATE.??? files. After selection of all applications for building CLIMATE.??? files from ASCIDATA.??? files, the software will construct them. As an aid, the reverse operation is also provided. By selecting *Tools / Build ASCII Data File...*, the user invokes the **Pick Applications** dialog box, which is used in a similar fashion. This enables the user easily and quickly to build sample ASCIDATA.??? files, created from the example

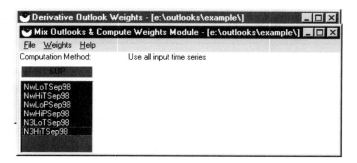

Figure A2-9. Mix Outlooks & Compute Weights Module window.

CLIMATE.??? files supplied with the software. The user can then use them as a formatting guide when writing the ASCIDATA.??? files for his or her own application areas.

Mix Outlooks & Compute Weights Module

The third module callable from the main window of **Derivative Outlook Weights** is the **Mix Outlooks & Compute Weights Module** and is invoked through the "Weights" button or by selecting *Modules / Weights*. An example display is given in Figure A2-9. The **Mix Outlooks & Compute Weights Module** also displays the application directory in its title. This module allows the user to set parameters for, and to compute, the weights to be applied to the operational hydrology sample defined with the second module (**Simulation Settings Module**), by using the meteorology outlooks defined with the first module (**Climate Outlooks Settings Module**). Therefore, this module should be called only after desired pertinent information has been set and saved in the other two modules. More specifically, this module allows the user to pick meteorology outlooks from one or more application areas and mix them into a new overall priority ordering. It allows selection of the optimization technique and objective function to be used in searching the probability statements (representing all meteorological outlooks) for a solution. It also then allows the computation of the weights. After all settings have been defined in the **Simulation Settings Module** and the **Climate Outlooks Settings Module**, then the **Mix Outlooks & Compute Weights Module** may be invoked and run any number of times to compute alternative sets of weights as the user desires. Current values (if any) for the computation method, application area(s), and meteorological outlooks are displayed in the **Mix Outlooks & Compute Weights Module** window (see Figure A2-9). They appear in fields as follows. "Computation Method" displays the optimization method. If the method displayed is "Use linear programming solution," then the Simplex method solution of a set of linear equations is used with an objective function defined by the user. If the method displayed is "Use all input time series" (as shown in Figure A2-9) or "Use most climatic outlooks," then either procedural algorithm 1 or 2, respectively, is used with the minimization of the sum of squared differences between the weights and unity. The three-character mnemonic for each application area appears in the main window immediately above a list of the meteorological outlooks to be used in that area. Figure A2-9 shows only one application area ("SUP"). The choice of computation method, selection of application area(s) for use in computing the weights, definition of objective function to be used in the optimization solution (for a linear programming solution), and the final mixing (priority ordering) of

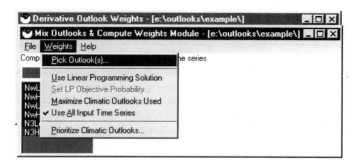

Figure A2-10. Mix Outlooks & Compute Weights Module Weights menu.

meteorology outlooks across application areas may be set or reset only in the **Mix Outlooks & Compute Weights Module**.

<u>Selecting Application Areas</u>
By selecting *Weights / Pick Outlook(s)...* , shown in Figure A2-10, the user invokes the **Pick Applications** dialog box (not shown). From there the user can select, by check-marking, one or more (if available) application areas in the standard Windows fashion. The application areas listed in the dialog box consist of only those named and saved in the last invocation of **Simulation Settings Module** for which there exists a CLIMATE.??? file (??? = the application mnemonic) and for which there exists a CLIMOTLK.??? file (created and saved in the **Climate Outlooks Settings Module** for each application). For each application area selected in the **Pick Applications** dialog box, upon exit a corresponding column will appear in the main window of the **Mix Outlooks & Compute Weights Module**.

<u>Selecting Meteorology Outlooks</u>
Each column in the main window of the **Mix Outlooks & Compute Weights Module** represents an application area with a list of the meteorology outlooks defined for that area (in the **Climate Outlooks Settings Module**). The user may then define which of the outlooks in the list are to be used in the computation of weights by selecting them in the standard Windows fashion. After doing this for all of the application areas (columns) in the main window, the user may then mix and prioritize them. Note that a total of only 100 outlooks may be selected; if more are selected at this point, the mixing and prioritizing step will not be allowed (and hence the computation of weights will not be allowed), until 100 or fewer are selected.

<u>Mixing and Prioritizing Meteorology Outlooks</u>
By selecting *Weights / Prioritize Climatic Outlooks...* , shown in Figure A2-10, the user invokes the **Order by Priority (by dragging)** dialog box (not shown) for use in arbitrarily ordering the probability statements shown therein. Then the user may simply drag the statements, displayed there, in standard Windows fashion to reorder and mix them. This is very similar to the action described previously for the **Order by Priority (by dragging)** dialog box in the **Climate Outlooks Settings Module**. The statements listed here, however, contain identifiers as to their respective application areas.

Selecting Optimization Methodology

By check-marking *Weights / Use Linear Programming Solution*, shown in Figure A2-10, the user enables the use of linear programming and also enables the next menu entry, *Set LP Objective Probability...* (otherwise grayed out). By clearing the check mark on *Weights / Use Linear Programming Solution*, the user enables minimization of the sum of squares of differences between the weights and unity and also enables menu entries for *Maximize Climatic Outlooks Used* and *Use All Input Time Series* (otherwise grayed out). Simply clicking with the mouse in the "Computation Method" field in the **Mix Outlooks & Compute Weights Module** window also alternates between the three choices ("Use linear programming solution," "Use most climatic outlooks," and "Use all input time series").

Selecting Linear Programming Objective Function

If linear programming is to be used and the menu item for *Set LP Objective Probability...* has been enabled, then its selection will invoke the **Define LP Objective Function Events** dialog box (not shown). With this tool, the user can define any event or combination of events. Any existing user-defined events are displayed and may be added to by selecting the middle button ("define new event"). When selected, the **Define LP Objective Function Event** dialog box (not shown) is invoked to actually set the event (note that the former is plural and the latter is singular). The user can then enter relevant variable type (temperature or precipitation), application area name (choices are among all application areas represented in the application directory with CLIMATE.??? files), reference quantile definitions, and event dates. The beginning and ending dates may be any day from January 1, 1995, through December 31 of the year 2 years after the current year. (The 2-year period is changeable by selecting *File / Reference Settings / User Event Minimum Period...* , which invokes the **Enter A Number** dialog box, not shown.) The user-defined events may be further combined in the parent **Define LP Objective Function Events** dialog box by selecting one or more in the following fashion. After selecting the first event by clicking on it with the mouse, the user may add others to it such that the result represents the intersection (AND) of all the events by *shift*-clicking on them. After selecting the first event by clicking on it with the mouse, others may be added to it such that the result represents the union (OR) of all the events by *control*-clicking on them. By *alt*-clicking on any single event, its complement is defined. By using these actions to define events and their intersections, unions, and complements, the user can represent any complex event. By right-clicking on any of the defined events, the objective function for the optimization is defined to maximize the probability of that event.

Once all aspects of the mixed meteorological outlooks for all application area(s) are defined, the optimization selected, and the objective function selected, the weights may be computed by selecting *File / Calculate Weights*. This action first invokes the **Order by Priority (by dragging)** dialog box (not shown), for the user to double-check his or her mixing and prioritizing of the probability statements. Then it computes the weights and places them into the file, OTLKWGTS.$$$, along with an indication of which statements were satisfied; it also places all probability statements, written in terms of their inclusion coefficients [see Equations (6-10), (6-11), (7-5), (8-11), or (10-15)], into the file EQUATION.$$$.

Exercise A2-4

As a continuing exercise, run the *Derivative Outlook Weights* program or, if it is already running, exit from any module currently running, if necessary, to return to the **Derivative Outlook Weights** main window. Select *Application / Set Up Example...* to invoke the **Pick An Example** dialog box. Select "Example04," and click on the "accept selection" button on the left (let the cursor rest on any of the buttons for a hint to appear as to its function). (This action clears the EXAMPLE subdirectory; extracts files into it from the EXAMPLE04.ZIP archive, located in the EXAMPLES DATA subdirectory; and informs the user of its successful completion.) The main window should appear similar to that shown in Figure A2-2. Then open the **Mix Outlooks & Compute Weights Module** by selecting either the "Weights" button or *Modules / Weights*. The window display should be blank. Select *Weights / Pick Outlook(s)...* , to invoke the **Pick Applications** dialog box, and select the "SUP" application, which is the only one shown. Click the "accept selection" button (left most button). The window should no longer be blank, but should show the continuing example as in Figure A2-9. Click with the mouse in the field "Computation Method" until it reads "Use all input time series" or (alternatively) select *Weights / Use All Input Time Series*. (If menus are used and the menu item *Weights / Use All Input Time Series* is grayed out, then clear the check mark on *Weights / Use Linear Programming Solution*. This will enable the desired menu item so that it then may be selected.) In the column in the main window of the **Mix Outlooks & Compute Weights Module** entitled "SUP" (which should be the only column present), select all of the meteorology outlooks listed there, if not already selected. Select *Weights / Prioritize Climatic Outlooks...* to invoke the **Order by Priority (by dragging)** dialog box. Experiment with the order, but make sure to reset priorities to initial with the third button from the left before accepting the ordering. (Note that reordering is redundant in this simple example since only one application area, SUP, is present and its meteorology equations were already ordered in the **Climate Outlooks Settings Module**.) Select *File / Calculate Weights* to first invoke (automatically) the **Order by Priority (by dragging)** dialog box (simply accept the settings this time) and then to calculate the weights. Inspect the file EQUATION.$$$ in the application directory (subdirectory EXAMPLE) and compare with Equation (7-15).

NOTES

- The user can perform the steps in almost any order after selecting an application directory. Of course, the weights computation must be the last step.

- *Help / Instructions...* , in any of the modules, gives context-sensitive suggestions depending upon the actions of the user. In the **Simulation Settings Module**, if the current directory is invalid (contains no appropriate CLIMATE.??? file) or there is a file read error on the CLIMATE.??? file, then optional instructions are available on the required format of the CLIMATE.??? file.

- Invalid responses to any query will not be accepted; each query (dialog box) will remain until answered validly or canceled.

- Defaults for many queries (dialog boxes) are used extensively and are determined by the last answers to the queries in a current or preceding session.

- The user can change settings or move or resize program windows and dialog boxes; the program saves the changes automatically. It also saves all user-entered climate outlook information automatically.

- In the **Climate Outlooks Settings Module**, the user can change display options by selecting _File / Reference Settings / Information Display_ or _File / Reference Settings / Display-2_ or _File / Reference Settings / Display-3_. In the **Simulation Settings Module**, the user can change display options by selecting _File / Reference Settings / Information Display_.

- The program allows only one invocation of itself for safety's sake; if invoked while already running, the already-running instance is made the active window.

FILES

The program creates various files so that it may keep track of information. These files are created both in the installation directory and in the user's application area directories (e.g., subdirectory EXAMPLE). Many of these files are of no interest to the user and, furthermore, the user should not disturb them. Some files that may be of interest are detailed here in alphabetical order.

- CLIMATE.??? is a historical data file supplied by the user. It is a direct access file with two 2-byte integers per record. Record 1 is the Julian day and year of the start date of the historical data; record 2 is the Julian day and year of the end date of the historical data. The remaining records contain air temperature (first) and precipitation (second) in each record for each day of the historical period between the two dates (inclusive), listed chronologically. _Any units can be used; no missing data are allowed._ Note: The user can build this file from an ASCII data file by selecting _Tools / Build CLIMATE.??? File..._ in the program's **Simulation Settings Module**; see the subsection entitled "User-Supplied Historical Climate Data." The user can build an example ASCII data file from the example CLIMATE.??? files, supplied with the software, by selecting _Tools / Build ASCII Data File..._ in the **Simulation Settings Module**; see the subsection entitled "User-Supplied Historical Climate Data." This allows inspection of the required format to use in the ASCIDATA.??? files.

- CLIMOTLK.??? is an output file created by the program's **Climate Outlooks Settings Module** for use in the **Mix Outlooks & Compute Weights Module**. It contains the climate outlook information selected by the user: climatic outlook start dates, number of climatic outlook probability statements to be used, information on each climatic outlook (period, meteorological variable, and probability information), priority listing, reference period dates, and quantile periods.

- EQUATION.$$$ is an output file created by the program's **Mix Outlooks & Compute Weights Module**. It contains the equation coefficients, converted from the user's selected climate outlook information for all application areas, as in Equation (6-11) and the constraint equations in Equations (7-5b), the constraint equations in Equations (8-11c) and (8-11d), or both the objective function and constraint equations in Equations (10-15).

- OTLKEC1.??? is an internal file created by the program's **Climate Outlooks Settings Module**. It contains user-input values from EC's 1-month forecasts of most-probable air temperature events in an expanding database.

- OTLKEC3X.??? is an internal file created by the program's **Climate Outlooks Settings Module**. It contains user-input values from EC's four 3-month seasonal forecasts of most-probable air temperature and precipitation events in an expanding database. Earlier versions were named OTLKEC3.??? and may be converted to the present format by selecting _Tools / Convert Old _E_C 3-m Data..._ in the **Climate Outlooks Settings Module** (see the section Converting Old Forecast Databases).

- OTLKN610.??? is an internal file created by the program's **Climate Outlooks Settings Module**. It contains user-input values from NOAA's day 6 through day 10 (5-day) forecasts of most-probable air temperature and precipitation events in an expanding database.

- OTLKN814.??? is an internal file created by the program's **Climate Outlooks Settings Module**. It contains user-input values from NOAA's day 8 through day 14 (second-week) forecasts of air temperature and precipitation probabilities in an expanding database.

- OTLKSETS.??? is an internal file created by the program's **Climate Outlooks Settings Module**. It contains user-input values from NOAA's _Climate Outlook_ of one 1-month and thirteen 3-month forecasts of air temperature and precipitation probabilities in an expanding database. Earlier versions of OTLKSETS.??? (188-byte line lengths) may be converted to the present format (493-byte line lengths) by selecting _Tools / Convert Old _N_OAA 1-3m Data..._ in the **Climate Outlooks Settings Module**. Execution with a file already in the new format will not change the file (see the section Converting Old Forecast Databases).

- OTLKUSER.??? is an internal file created by the program's **Climate Outlooks Settings Module**. It contains user-input values for probabilistic meteorology forecasts of the user's own definition, in an expanding database.

- OTLKWGTS.$$$ is an output file created by the program's **Mix Outlooks & Compute Weights Module**. It contains the outlook weights corresponding to the user's settings and an indication of which meteorology probability forecast statements were used. Those marked with a plus (+) were satisfied, and those marked with a minus (−) were not.

- PRIORITY.??? is an internal file created by the program. It contains user-input ordering of the climate outlooks for application area, ???, last used by the program.

- SIMULATE.$$$ is an output file created by the program's **Simulation Settings Module** for use in the **Mix Outlooks & Compute Weights Module**. It contains user-selected forecast information: application mnemonics, forecast start date, historical start date, date forecast is made, forecast length, number of time series (historical record pieces) to be used in the forecast, and start year for each piece.

ADDITIONAL EXERCISES
The examples contained in Chapters 7, 8, 9, and 10 are detailed here. Shortcuts are provided to data entry in the form of archived databases.

Exercise A2-5
This exercise uses the same data as Exercise A2-4 and so is set up as described in Exercise A2-4. After the user has performed that exercise, the "SUP" application should already be selected in the **Mix Outlooks & Compute Weights Module**, and all of the meteorological outlooks listed in the "SUP" column there should also be selected, as in Figure A2-9. Also, the field "Computation Method" should read "Use all input time series." Now, select _Weights / _P_rioritize Climatic Outlooks..._ to invoke the **Order by Pri-**

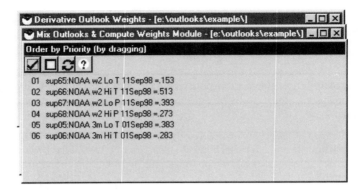

Figure A2-11. Priority order in Mix Outlooks & Compute Weights Module.

ority (by dragging) dialog box. Make sure that the six statements match the priority listing in Equations (7-13), as shown for the **Order by Priority (by dragging)** dialog box in Figure A2-11, and accept the ordering. [Click on the second button from the left ("toggle equation / info display") to place the display in the mode shown, if necessary.] Select _File / Calculate Weights_ to first invoke (automatically) the **Order by Priority (by dragging)** dialog box (simply accept the settings this time) and then to calculate the weights. Inspect the file EQUATION.$$$ in the application area directory (subdirectory EXAMPLE) and compare with Equation (7-15), as in Exercise A2-4. They should be identical. Inspect the file OTLKWGTS.$$$ in the application directory (subdirectory EXAMPLE) and compare the weights there with Table 7-5. They should be identical. Note the second line in the OTLKWGTS.$$$ file. It lists the probability statements by name only in the same order as selected, but each is followed with a plus (+) or minus (−) sign, indicating that it was included or excluded, respectively, in the computation of the weights. The last two statements in this list are indicated as not used.

Now, change the computation method (back in the **Mix Outlooks & Compute Weights Module**) to use the most climatic outlooks by clicking on the "Computation Method" field until it reads "Use most climatic outlooks" or by selecting _Weights / Maximize Climatic Outlooks Used_. While using the same priorities as pictured in Equations (7-13) and Figure A2-11 (as before), select _File / Calculate Weights_ to first invoke (automatically) the **Order by Priority (by dragging)** dialog box (simply accept the settings this time) and then to calculate the weights. Inspect the file OTLKWGTS.$$$ in the application directory (subdirectory EXAMPLE) and compare the weights there with Table 7-6. They should be identical. Note that all probability statements, listed in the second line in the OTLKWGTS.$$$ file, are indicated as used, but that there are zero weights for years 1953, 1956, and 1961 (indices 6, 9, and 14, respectively, in Table 7-6).

Exercise A2-6
As a continuing exercise, run the _Derivative Outlook Weights_ program or, if it is already running, exit from any module currently running, if necessary, to return to the **Derivative Outlook Weights** main window. Either continue working with the data in the EXAMPLE subdirectory left at the completion of Exercise A2-4 or A2-5, or select _Application / Set Up Example..._ and select "Example06" to clear the EXAMPLE subdirectory and to extract files into it from the EXAMPLE06.ZIP archive. In the former case, the user will have to adjust settings in the various modules to match the windows in Figures A2-

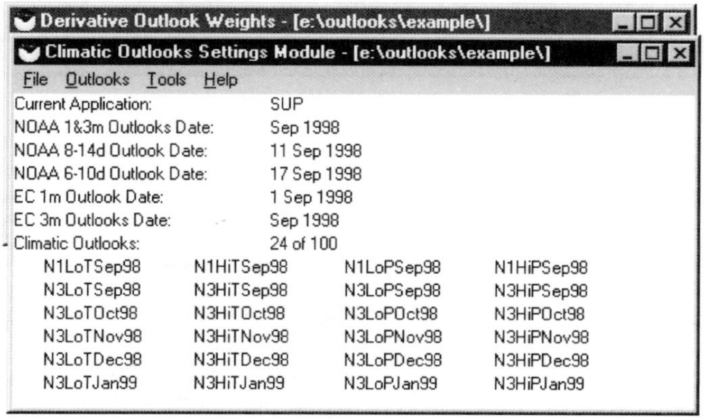

Figure A2-12. Climatic outlooks settings for Exercise A2-6.

12 through A2-16. See the earlier instructions and exercises in this appendix for information on making these settings. In the latter case, the settings should already match the windows in Figures A2-12 through A2-16, with the possible exception of the "Computation Method" field in Figure A2-15. If the settings there are different, click that field until

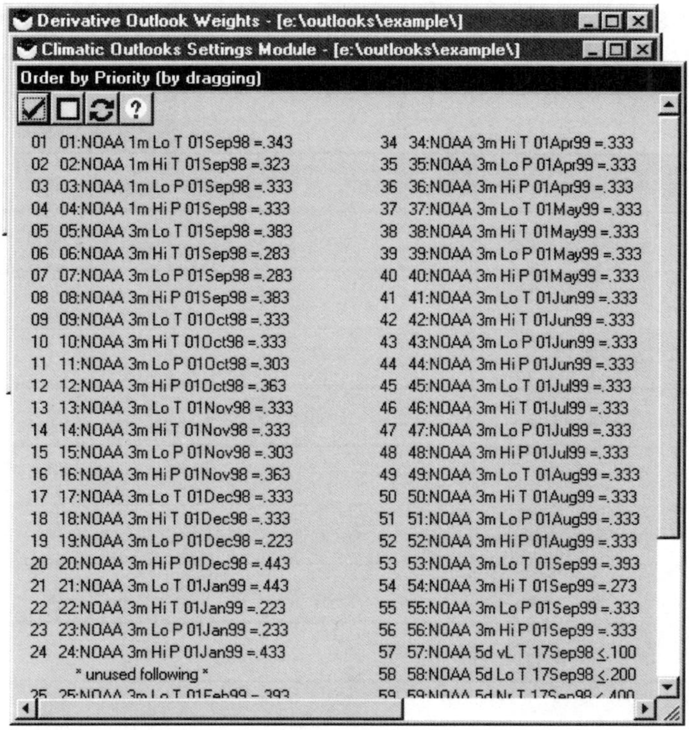

Figure A2-13. Climatic outlooks priority order for Exercise A2-6.

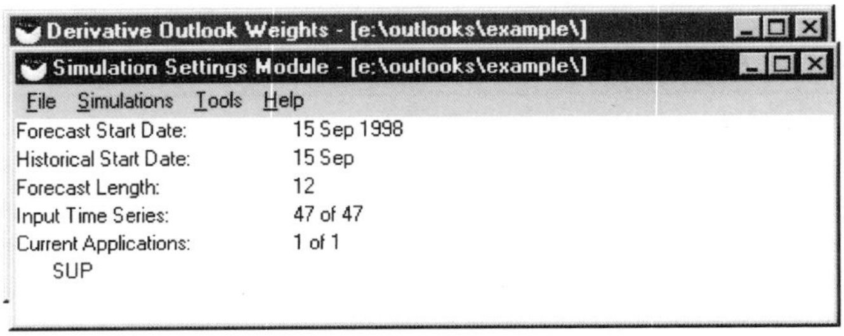

Figure A2-14. Simulation settings for Exercises A2-6, A2-7, and A2-8.

it agrees with Figure A2-15. In both Figures A2-13 and A2-16, the user may have to click on the second button from the left ("toggle equation / info display") to place the display in the mode shown there. (Note that Figures A2-13 and A2-16 are redundant in this example; both show the same ordering.) In the **Mix Outlooks & Compute Weights Module**, select *File / Calculate Weights* to first invoke (automatically) the **Order by Priority (by dragging)** dialog box (simply accept the settings again) and then to calculate the weights. Inspect the file EQUATION.$$$ in the application directory (subdirectory EXAMPLE) and compare with Figure 7-3. They should be identical. Inspect the file OTLKWGTS.$$$ and compare the weights there with Table 7-13. They should be identical. Note that all probability statements, listed in the second line in the OTLKWGTS.$$$ file, are indicated as used, but that there are 12 zero weights, in agreement with Table 7-13.

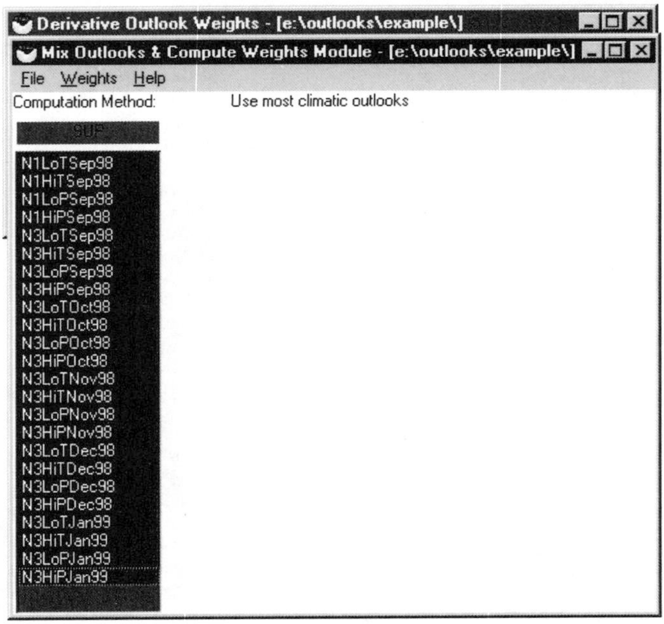

Figure A2-15. Weight computation settings for Exercise A2-6.

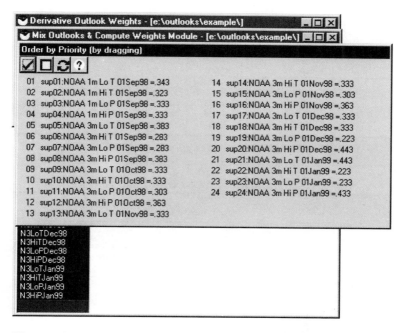

Figure A2-16. Weight computation priority order for Exercise A2-6.

Exercise A2-7

As a continuing exercise, run the *Derivative Outlook Weights* program or, if it is already running, exit from any module currently running, if necessary, to return to the **Derivative Outlook Weights** main window. Select *Application / Set Up Example...* and select "Example07" to clear the EXAMPLE subdirectory and to extract files into it from the EXAMPLE07.ZIP archive. Note that the included agency outlooks have been redefined. In the **Climate Outlooks Settings Module**, select *File / Reference Settings / Change Agency Outlooks Used...* to invoke the **Change Agency Outlooks Used** dialog box and inspect the settings. All agency outlooks have been eliminated so that there are now 100 user-defined equations. Do not change anything here, but accept the settings. (If the user has changed anything, restore the settings by clicking the "reset to initial" button, the third button from the left.) To further inspect the user-defined equations, while still in the **Climate Outlooks Settings Module** select *Outlooks / Set User-Definable Probability Statement...* to invoke the **Define User Probability Statements** dialog box and compare with Equations (4-9). By clearing and then re-check-marking any equation there, the user has the opportunity to inspect or change the equation. Try one (clear and re-check-mark an equation) to invoke the **Define User Probability Statement** dialog box, and note that the reference period used is 1961–90. Exit from these dialog boxes by selecting the "cancel" button in each, and then exit from the **Climate Outlooks Settings Module**.

The settings elsewhere should match the windows in Figures A2-14 and A2-17 through A2-20, with the possible exception of the "Computation Method" field in Figure A2-18 (in the **Mix Outlooks & Compute Weights Module**). If the setting there is different, click that field until it agrees with Figure A2-18. In Figures A2-19 and A2-20, the user may have to click on the second button from the left ("toggle equation / info

183

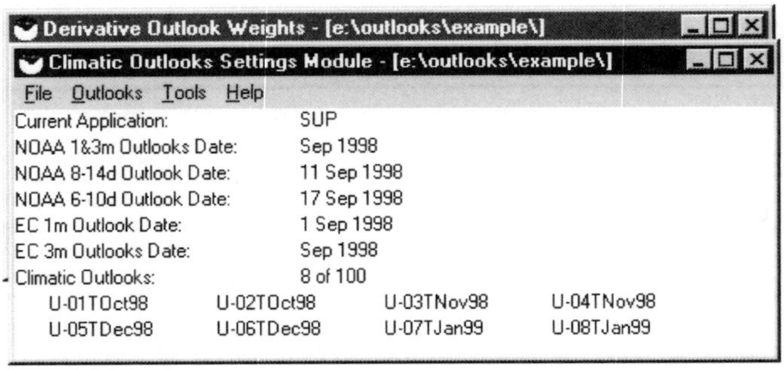

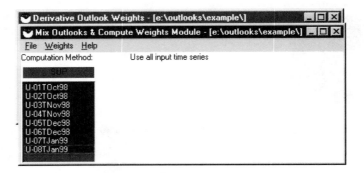

Figure A2-17. Climatic outlooks settings for Exercise A2-7.

Figure A2-18. Weight computation settings for Exercise A2-7.

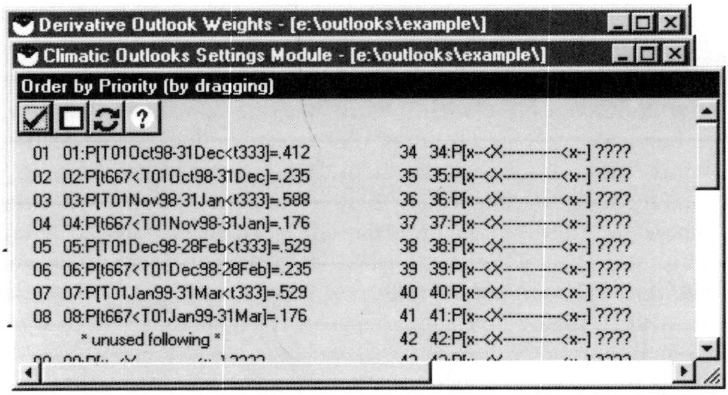

Figure A2-19. Climatic outlooks priority order for Exercise A2-7.

display") to place the display in the mode shown there. (Note that these windows are redundant in this example; both show the same ordering.) In the **Mix Outlooks & Compute Weights Module**, select *File / Calculate Weights* to first invoke (automatically) the **Order by Priority (by dragging)** dialog box (simply accept the settings again) and then to calculate the weights. Inspect the file EQUATION.$$$ in the ap-

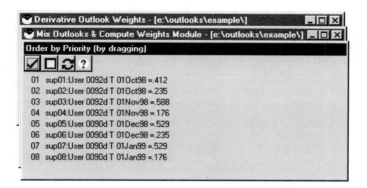

Figure A2-20. Weight computation priority order for Exercise A2-7.

plication directory (subdirectory EXAMPLE) and compare with Equation (7-16). They should be identical. Inspect the file OTLKWGTS.$$$ and compare the weights there with Table 7-15. They should be identical. Note that the first seven of the eight probability statements, listed in the second line in the OTLKWGTS.$$$ file, are indicated as used.

Now, change the computation method to use the most climatic outlooks by clicking on the "Computation Method" field until it reads "Use most climatic outlooks" or by selecting *Weights / Maximize Climatic Outlooks Used*. (If menus are used and menu item *Weights / Maximize Climatic Outlooks Used* is grayed out, then clear the check mark on *Weights / Use Linear Programming Solution*. This will enable the desired menu item so that it may then be selected.) While using the same priorities, select *File / Calculate Weights* to first invoke (automatically) the **Order by Priority (by dragging)** dialog box (simply accept the settings this time) and then to calculate the weights. Inspect the OTLKWGTS.$$$ file in the application directory (subdirectory EXAMPLE) and compare the weights there with Table 7-16. They should be identical. Note that all probability statements, listed in the second line in the OTLKWGTS.$$$ file, are indicated as used, but that there are 8 zero weights in Table 7-16.

Exercise A2-8

As a continuing exercise, run the *Derivative Outlook Weights* program or, if it is already running, exit from any module currently running, if necessary, to return to the **Derivative Outlook Weights** main window. Select *Application / Set Up Example...* and select "Example08" to clear the EXAMPLE subdirectory and to extract files into it from the EXAMPLE08.ZIP archive. The settings should match the windows in Figures A2-14 and A2-21 through A2-24, with the possible exception of the "Computation Method" field in Figure A2-23. If the settings there are different, click that field until it agrees with Figure A2-23. In both Figures A2-22 and A2-24, the user may have to click on the second button from the left ("toggle equation / info display") to place the display in the mode shown there. (Note that Figures A2-22 and A2-24 are redundant in this example; both show the same ordering, as given in Figure 8-4.) In the **Mix Outlooks & Compute Weights Module**, select *File / Calculate Weights* to first invoke (automatically) the **Order by Priority (by dragging)** dialog box (simply accept the settings again) and then to calculate the weights. Inspect the file EQUATION.$$$ in the application directory (subdirectory EXAMPLE) and compare with Figure 8-5. They should be identical. Inspect the file OTLKWGTS.$$$ and compare the weights there with Table 8-1. They

Derivative Outlook Weights - [e:\outlooks\example\]

Climatic Outlooks Settings Module - [e:\outlooks\example\]

File Outlooks Tools Help

Current Application:	SUP
NOAA 1&3m Outlooks Date:	Sep 1998
NOAA 8-14d Outlook Date:	11 Sep 1998
NOAA 6-10d Outlook Date:	17 Sep 1998
EC 1m Outlook Date:	1 Sep 1998
EC 3m Outlooks Date:	Sep 1998
Climatic Outlooks:	41 of 100

NwLoTSep98	NwHiTSep98	NwLoPSep98	NwHiPSep98
N5vLTSep98	N5LoTSep98	N5NrTSep98	N5HiTSep98
N5vHTSep98	N5LoPSep98	N5NrPSep98	N5HiPSep98
N1LoTSep98	N1HiTSep98	E1LoTSep98	E1HiTSep98
N1LoPSep98	N1HiPSep98	N3LoTSep98	N3HiTSep98
N3LoPSep98	N3HiPSep98	E3LoPSep98	E3NrPSep98
E3HiPSep98	N3LoTOct98	N3HiTOct98	N3LoPOct98
N3HiPOct98	N3LoTNov98	N3HiTNov98	N3LoPNov98
N3HiPNov98	N3LoTDec98	N3HiTDec98	N3LoPDec98
N3HiPDec98	N3LoTJan99	N3HiTJan99	N3LoPJan99
N3HiPJan99			

Figure A2-21. Climatic outlooks settings for Exercise A2-8.

should be identical. Note that in the second line in the OTLKWGTS.$$$ file, all probability statements except the last 14 are indicated as used, and that there are 12 zero weights, in agreement with Table 8-1.

Exercise A2-9
As a continuing exercise, run the *Derivative Outlook Weights* program or, if it is already running, exit from any module currently running, if necessary, to return to the **Derivative Outlook Weights** main window. Select *Application / Set Up Example…* and select "Example09" to clear the EXAMPLE subdirectory and to extract files into it from the EXAMPLE09.ZIP. The **Simulation Settings Module** settings should match Figure A2-25 (with the last 41 of the possible 47 input time series selected and the five application areas as indicated), and the **Climate Outlooks Settings Module** settings should agree with Figure A2-26. That is, even though Figure A2-26 shows the **Mix Outlooks & Compute Weights Module**, the **Climate Outlooks Settings Module** should be employed on each of the five application areas (SUP, MIC, HUR, GEO, and ERI) to select and order the eight equations for each area, as identified in Figure A2-26, respectively. The actual form of each equation can be seen in the **Climate Outlooks Settings Module** to agree with Figure 9-4. In the **Mix Outlooks & Compute Weights Module**, make sure that the bottom four equations of each set of eight are highlighted for each application area as shown in Figure A2-26. Also ensure that the priority order is as shown in Figure A2-27, in agreement also with the ordering in Figure 9-4. The user may have to click on the second button from the left ("toggle equation / info display") to place the display in the mode shown there. Finally, check the "Computation Method" field with Figure A2-26. If the settings are different, click that field until it agrees with Figure A2-26. Note that this is the first exercise for multiple application areas, and there is no longer a redundancy between the priority ordering in the (single-area) application within the **Climate Outlooks Settings Module** and that in the **Mix Outlooks & Compute Weights Module**. With multiple application areas, the priority ordering in the **Mix Outlooks & Compute Weights Module** becomes impor-

Derivative Outlook Weights - [e:\outlooks\example\]

Climatic Outlooks Settings Module - [e:\outlooks\example\]

Order by Priority (by dragging)

01	65:NOAA w2 Lo T 11Sep98 =.153	35	18:NOAA 3m Hi T 01Dec98 =.333
02	66:NOAA w2 Hi T 11Sep98 =.513	36	19:NOAA 3m Lo P 01Dec98 =.223
03	67:NOAA w2 Lo P 11Sep98 =.393	37	20:NOAA 3m Hi P 01Dec98 =.443
04	68:NOAA w2 Hi P 11Sep98 =.273	38	21:NOAA 3m Lo T 01Jan99 =.443
05	57:NOAA 5d vL T 17Sep98 ≤.100	39	22:NOAA 3m Hi T 01Jan99 =.223
06	58:NOAA 5d Lo T 17Sep98 ≤.200	40	23:NOAA 3m Lo P 01Jan99 =.233
07	59:NOAA 5d Nr T 17Sep98 ≤.400	41	24:NOAA 3m Hi P 01Jan99 =.433
08	60:NOAA 5d Hi T 17Sep98 >.200		* unused following *
09	61:NOAA 5d vH T 17Sep98 ≤.100	42	25:NOAA 3m Lo T 01Feb99 =.393
10	62:NOAA 5d Lo P 17Sep98 ≤.333	43	26:NOAA 3m Hi T 01Feb99 =.273
11	63:NOAA 5d Nr P 17Sep98 >.334	44	27:NOAA 3m Lo P 01Feb99 =.323
12	64:NOAA 5d Hi P 17Sep98 ≤.333	45	28:NOAA 3m Hi P 01Feb99 =.343
13	01:NOAA 1m Lo T 01Sep98 =.343	46	29:NOAA 3m Lo T 01Mar99 =.403
14	02:NOAA 1m Hi T 01Sep98 =.323	47	30:NOAA 3m Hi T 01Mar99 =.263
15	69:EnCa 1m Lo T 01Sep98 >.333	48	31:NOAA 3m Lo P 01Mar99 =.333
16	71:EnCa 1m Hi T 01Sep98 ≤.333	49	32:NOAA 3m Hi P 01Mar99 =.333
17	03:NOAA 1m Lo P 01Sep98 =.333	50	33:NOAA 3m Lo T 01Apr99 =.333
18	04:NOAA 1m Hi P 01Sep98 =.333	51	34:NOAA 3m Hi T 01Apr99 =.333
19	05:NOAA 3m Lo T 01Sep98 =.383	52	35:NOAA 3m Lo P 01Apr99 =.333
20	06:NOAA 3m Hi T 01Sep98 =.283	53	36:NOAA 3m Hi P 01Apr99 =.333
21	07:NOAA 3m Lo P 01Sep98 =.283	54	37:NOAA 3m Lo T 01May99 =.333
22	08:NOAA 3m Hi P 01Sep98 =.383	55	38:NOAA 3m Hi T 01May99 =.333
23	75:EnCa 3m Lo P 01Sep98 ≤.333	56	39:NOAA 3m Lo P 01May99 =.333
24	76:EnCa 3m Nr P 01Sep98 ≤.334	57	40:NOAA 3m Hi P 01May99 =.333
25	77:EnCa 3m Hi P 01Sep98 >.333	58	41:NOAA 3m Lo T 01Jun99 =.333
26	09:NOAA 3m Lo T 01Oct98 =.333	59	42:NOAA 3m Hi T 01Jun99 =.333
27	10:NOAA 3m Hi T 01Oct98 =.333	60	43:NOAA 3m Lo P 01Jun99 =.333
28	11:NOAA 3m Lo P 01Oct98 =.303	61	44:NOAA 3m Hi P 01Jun99 =.333
29	12:NOAA 3m Hi P 01Oct98 =.363	62	45:NOAA 3m Lo T 01Jul99 =.333
30	13:NOAA 3m Lo T 01Nov98 =.333	63	46:NOAA 3m Hi T 01Jul99 =.333
31	14:NOAA 3m Hi T 01Nov98 =.333	64	47:NOAA 3m Lo P 01Jul99 =.333
32	15:NOAA 3m Lo P 01Nov98 =.303	65	48:NOAA 3m Hi P 01Jul99 =.333
33	16:NOAA 3m Hi P 01Nov98 =.363	66	49:NOAA 3m Lo T 01Aug99 =.333
34	17:NOAA 3m Lo T 01Dec98 =.333	67	50:NOAA 3m Hi T 01Aug99 =.333

Figure A2-22. Climatic outlooks priority order for Exercise A2-8.

tant to establish priorities across *both* the different equations and the different application areas. In the **Mix Outlooks & Compute Weights Module**, select *File / Calculate Weights* to first invoke (automatically) the **Order by Priority (by dragging)** dialog box and then to calculate the weights. Inspect the file EQUATION.$$$ in the application directory (subdirectory EXAMPLE) and compare with Figure 9-5. They should be identical. Inspect the file OTLKWGTS.$$$ and compare the weights there with Table 9-5. They should be identical. Note that all probability statements, listed in the second line in the OTLKWGTS.$$$ file, are indicated as used, and that there are only 2 zero weights, in agreement with Table 9-5.

Exercise A2-10

This exercise uses the same data as Exercise A2-9. Therefore, run the *Derivative Outlook Weights* program or, if it is already running, exit from any module currently running, if necessary, to return to the **Derivative Outlook Weights** main window. Se-

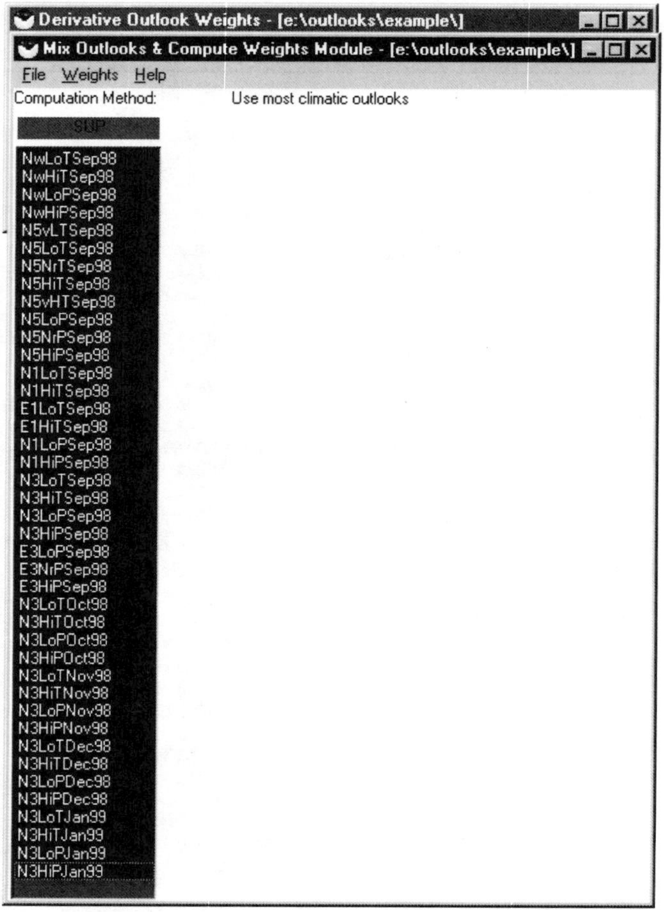

Figure A2-23. Weight computation settings for Exercise A2-8.

lect *Application / Set Up Example...* and select "Example09" (not "10") to clear the EXAMPLE subdirectory and to extract files into it from the EXAMPLE09.ZIP archive. The **Simulation Settings Module** settings should match Figure A2-25, and the **Climate Outlooks Settings Module** settings should agree with Figure A2-26, as in Exercise A2-9. That is, the **Climate Outlooks Settings Module** should be employed on each of the five application areas (SUP, MIC, HUR, GEO, and ERI) to select and order the eight equations for each area, as identified in Figure A2-26, respectively. (The actual form of each equation can be seen in the **Climate Outlooks Settings Module** to agree with Figure 9-4). As in Exercise A2-9, in the **Mix Outlooks & Compute Weights Module**, make sure that the bottom four equations of each set of eight are highlighted for each application area as shown in Figure A2-26. Also ensure that the priority order is as shown in Figure A2-27, in agreement also with the ordering in Figure 9-4. The user may have to click on the second button from the left ("toggle equation / info display") to place the display in the mode shown there. [Note that this is another

Mix Outlooks & Compute Weights Module - [e:\outlooks\example\]

Order by Priority (by dragging)

01 sup65:NOAA w2 Lo T 11Sep98 =.153	23 sup75:EnCa 3m Lo P 01Sep98 ≤.333
02 sup66:NOAA w2 Hi T 11Sep98 =.513	24 sup76:EnCa 3m Nr P 01Sep98 ≤.334
03 sup67:NOAA w2 Lo P 11Sep98 =.393	25 sup77:EnCa 3m Hi P 01Sep98 >.333
04 sup68:NOAA w2 Hi P 11Sep98 =.273	26 sup09:NOAA 3m Lo T 01Oct98 =.333
05 sup57:NOAA 5d vL T 17Sep98 ≤.100	27 sup10:NOAA 3m Hi T 01Oct98 =.333
06 sup58:NOAA 5d Lo T 17Sep98 ≤.200	28 sup11:NOAA 3m Lo P 01Oct98 =.303
07 sup59:NOAA 5d Nr T 17Sep98 ≤.400	29 sup12:NOAA 3m Hi P 01Oct98 =.363
08 sup60:NOAA 5d Hi T 17Sep98 >.200	30 sup13:NOAA 3m Lo T 01Nov98 =.333
09 sup61:NOAA 5d vH T 17Sep98 ≤.100	31 sup14:NOAA 3m Hi T 01Nov98 =.333
10 sup62:NOAA 5d Lo P 17Sep98 ≤.333	32 sup15:NOAA 3m Lo P 01Nov98 =.303
11 sup63:NOAA 5d Nr P 17Sep98 >.334	33 sup16:NOAA 3m Hi P 01Nov98 =.363
12 sup64:NOAA 5d Hi P 17Sep98 ≤.333	34 sup17:NOAA 3m Lo T 01Dec98 =.333
13 sup01:NOAA 1m Lo T 01Sep98 =.343	35 sup18:NOAA 3m Hi T 01Dec98 =.333
14 sup02:NOAA 1m Hi T 01Sep98 =.323	36 sup19:NOAA 3m Lo P 01Dec98 =.223
15 sup69:EnCa 1m Lo T 01Sep98 >.333	37 sup20:NOAA 3m Hi P 01Dec98 =.443
16 sup71:EnCa 1m Hi T 01Sep98 ≤.333	38 sup21:NOAA 3m Lo T 01Jan99 =.443
17 sup03:NOAA 1m Lo P 01Sep98 =.333	39 sup22:NOAA 3m Hi T 01Jan99 =.223
18 sup04:NOAA 1m Hi P 01Sep98 =.333	40 sup23:NOAA 3m Lo P 01Jan99 =.233
19 sup05:NOAA 3m Lo T 01Sep98 =.383	41 sup24:NOAA 3m Hi P 01Jan99 =.433
20 sup06:NOAA 3m Hi T 01Sep98 =.283	
21 sup07:NOAA 3m Lo P 01Sep98 =.283	
22 sup08:NOAA 3m Hi P 01Sep98 =.383	

Figure A2-24. Weight computation priority order for Exercise A2-8.

exercise for multiple application areas and there is no redundancy between the priority ordering in each (single-area) application within the **Climate Outlooks Settings Module** and that in the **Mix Outlooks & Compute Weights Module**.] Check the "Computation Method" field, but now ensure that it reads "Use linear programming solution." Finally, select *Weights / Set LP Objective Probability...* to invoke the **Define LP Objective Function Events** dialog box. Inspect the single event portrayed there. It should look like Figure A2-28. To try defining this event for yourself, click on the middle button ("define new event") to invoke the **Define LP Objective Function Event** dialog box. Make the settings in this box agree with Figure A2-29. Note that the button for setting the application area name("Select or Define Application Area") displays the title "ALL" to its right. This corresponds to the data file for the combined application areas for Superior, Michigan, Huron, Georgian Bay, and Erie. Click the left

Derivative Outlook Weights - [e:\outlooks\example\]

Simulation Settings Module - [e:\outlooks\example\]

File Simulations Tools Help

Forecast Start Date:	15 Sep 1998
Historical Start Date:	15 Sep
Forecast Length:	12
Input Time Series:	41 of 47
Current Applications:	5 of 5
ERI GEO HUR MIC SUP	

Figure A2-25. Simulation settings for Exercises A2-9 and A2-10.

Derivative Outlook Weights - [e:\outlooks\example\]

Mix Outlooks & Compute Weights Module - [e:\outlooks\example\]

File Weights Help

Computation Method: Use most climatic outlooks

LBR	GBD	HUR	MIC	SOR

N3LoTSep98	N3LoTSep98	N3LoTSep98	N3LoTSep98	N3LoTSep98
N3HiTSep98	N3HiTSep98	N3HiTSep98	N3HiTSep98	N3HiTSep98
N3LoPSep98	N3LoPSep98	N3LoPSep98	N3LoPSep98	N3LoPSep98
N3HiPSep98	N3HiPSep98	N3HiPSep98	N3HiPSep98	N3HiPSep98

Figure A2-26. Weight computation settings for Exercises A2-9 and A2-10.

Derivative Outlook Weights - [e:\outlooks\example\]

Mix Outlooks & Compute Weights Module - [e:\outlooks\example\]

Order by Priority (by dragging)

01	sup07:NOAA 3m Lo P 01Sep98 =.283	11	hur05:NOAA 3m Lo T 01Sep98 =.333
02	sup08:NOAA 3m Hi P 01Sep98 =.383	12	hur06:NOAA 3m Hi T 01Sep98 =.333
03	sup05:NOAA 3m Lo T 01Sep98 =.383	13	geo07:NOAA 3m Lo P 01Sep98 =.303
04	sup06:NOAA 3m Hi T 01Sep98 =.283	14	geo08:NOAA 3m Hi P 01Sep98 =.363
05	mic07:NOAA 3m Lo P 01Sep98 =.283	15	geo05:NOAA 3m Lo T 01Sep98 =.333
06	mic08:NOAA 3m Hi P 01Sep98 =.383	16	geo06:NOAA 3m Hi T 01Sep98 =.333
07	mic05:NOAA 3m Lo T 01Sep98 =.363	17	eri07:NOAA 3m Lo P 01Sep98 =.313
08	mic06:NOAA 3m Hi T 01Sep98 =.303	18	eri08:NOAA 3m Hi P 01Sep98 =.353
09	hur07:NOAA 3m Lo P 01Sep98 =.283	19	eri05:NOAA 3m Lo T 01Sep98 =.333
10	hur08:NOAA 3m Hi P 01Sep98 =.383	20	eri06:NOAA 3m Hi T 01Sep98 =.333

Figure A2-27. Weight computation priority order for Exercises A2-9 and A2-10.

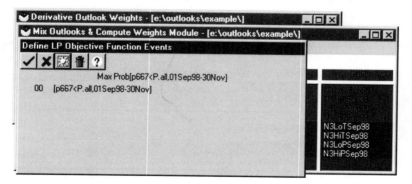

Figure A2-28. Objective function defining events for Exercise A2-10.

button ("OK") to save the results. (Make sure all settings are identical to Figure A2-29 if this event definition is used subsequently.) Now there are two event definitions in the **Define LP Objective Function Events** dialog box. Select one, by right-clicking on it with the mouse, as the defining event in the objective function and click on the left-

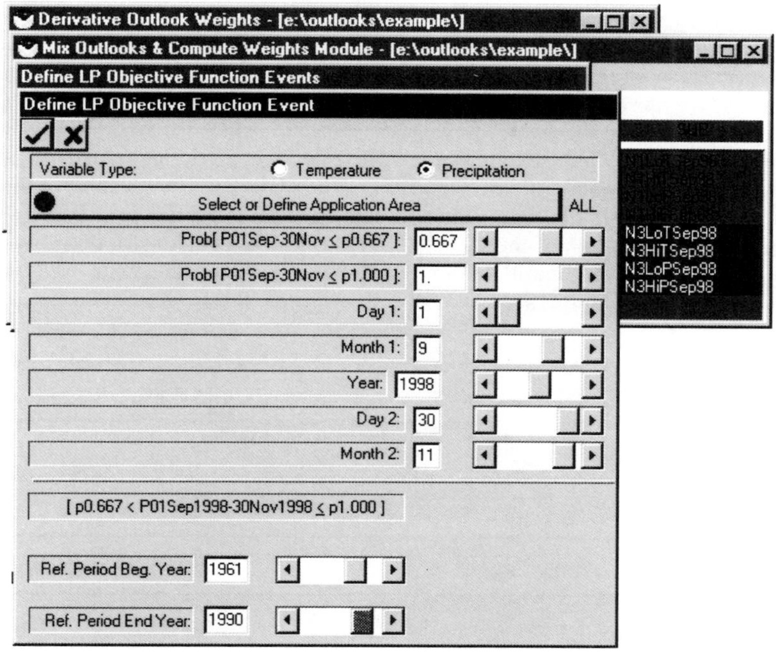

Figure A2-29. Defining event "00" for Exercise A2-10.

most button ("accept") to exit. In the **Mix Outlooks & Compute Weights Module**, select *File / Calculate Weights* to first invoke (automatically) the **Order by Priority (by dragging)** dialog box and then to calculate the weights. Inspect the file EQUATION.$$$ in the application directory (subdirectory EXAMPLE) and compare with Figure 10-2. They should be identical. Inspect the file OTLKWGTS.$$$ and compare the weights there with Table 10-1. They should be identical. Note that all probability statements, listed in the second line in the OTLKWGTS.$$$ file, are indicated as used, but there are several zero weights, in agreement with Table 10-1.

Exercise A2-11
This exercise uses the same data as Exercises A2-9 and A2-10. After performing Exercise A2-10, and while in the **Mix Outlooks & Compute Weights Module**, select *Weights / Set LP Objective Probability...* to invoke the **Define LP Objective Function Events** dialog box. It should look like Figure A2-28. Now, click on the middle button ("define new event") to invoke the **Define LP Objective Function Event** dialog box. Define an event as in the top line of Equation (10-6). Save by clicking on the left button ("OK") and repeat to define the event in the bottom line of Equation (10-6). [The combined SUP and MIC areas are represented by "SAM" when setting the application area name with the button ("Select or Define Application Area").] Back in the **Define LP Objective Function Events** dialog box, click on the first event just defined and shift-click on the second event just defined to define the intersection as a new event. Right-click on this newly created event (the intersection) to define it in the objective function. Then accept these changes with the leftmost button. When weights are computed, a unique optimum should be indicated. This is indicated

191

by the first line in the OTLKWGTS.$$$ file by the numbers "(1, 1)," indicating that there was only "1" unique optimum solution point found among the "1" nonunique optimum solutions searched.

Now, repeat the previous step to build the objective function in Equation (10-7). First, define each of the four events listed in the four lines of Equation (10-7). (The combined HUR, GEO, and ERI areas are represented by "HGE" in setting the application area name.) Then build their intersection. Then take the complement of the intersection and select this complement for defining the objective function. When weights are computed, the first line of the OTLKWGTS.$$$ file will contain "(5000, 7)," indicating that there were seven unique optimum solution points found among the first 5000 nonunique optimum solutions searched.

Appendix 3

MIN $\sum(w_i - 1)^2$ SUFFICIENT CONDITION

In the minimization of the sum of the squared deviations of each weight from unity, the constrained optimization of Equations (7-5) or (8-11) is converted into an unconstrained optimization through the introduction of the Lagrangian in Equation (7-6) or (8-16), respectively. Classical calculus is used to identify "critical points" where this function has zero first derivatives, as in Equations (7-7) or (8-17), representing necessary conditions for a minimization of Equation (7-6) or (8-16), respectively. However, critical points may also correspond to other zero-slope points such as maximum points or inflection points. Therefore, the critical point found by the solution of Equations (7-7) or (8-17) must be further checked to discern whether it represents a minimum. Such a check is called a *sufficient condition* and is developed here.

SUFFICIENCY FOR EQUATIONS (7-7)
Let

$$
\mathbf{x} = \begin{bmatrix} w_1 \\ \vdots \\ w_n \\ \lambda_1 \\ \vdots \\ \lambda_m \end{bmatrix}
\tag{A3-1}
$$

and

$$
\mathbf{x}' = \begin{bmatrix} w_1 & \cdots & w_n & \lambda_1 & \cdots & \lambda_m \end{bmatrix}
\tag{A3-2}
$$

where the bold notation denotes vectors. Equations (7-7) are written in vector notation as

$$
\frac{\partial L}{\partial \mathbf{x}} = \begin{bmatrix} \partial L/\partial w_1 \\ \vdots \\ \partial L/\partial w_n \\ \partial L/\partial \lambda_1 \\ \vdots \\ \partial L/\partial \lambda_m \end{bmatrix} = \begin{bmatrix} 2(w_1 - 1) - \sum_{k=1}^{m} \lambda_k \alpha_{k,1} \\ \vdots \\ 2(w_n - 1) - \sum_{k=1}^{m} \lambda_k \alpha_{k,n} \\ -\sum_{i=1}^{n} \alpha_{1,i} w_i + e_1 \\ \vdots \\ -\sum_{i=1}^{n} \alpha_{m,i} w_i + e_m \end{bmatrix} = \mathbf{0}
\tag{A3-3}
$$

where $\mathbf{0}$ denotes a column vector of all zeros. The second derivative matrix is

$$\frac{\partial(\partial L/\partial \mathbf{x})}{\partial \mathbf{x}'} = \begin{bmatrix} 2 & 0 & \cdots & 0 & -\alpha_{1,1} & -\alpha_{2,1} & \cdots & -\alpha_{m,1} \\ 0 & 2 & \cdots & 0 & -\alpha_{1,2} & -\alpha_{2,2} & \cdots & -\alpha_{m,2} \\ \vdots & \vdots & & \vdots & \vdots & \vdots & & \vdots \\ 0 & 0 & \cdots & 2 & -\alpha_{1,n} & -\alpha_{2,n} & \cdots & -\alpha_{m,n} \\ -\alpha_{1,1} & -\alpha_{1,2} & \cdots & -\alpha_{1,n} & 0 & 0 & \cdots & 0 \\ -\alpha_{2,1} & -\alpha_{2,2} & \cdots & -\alpha_{2,n} & 0 & 0 & \cdots & 0 \\ \vdots & \vdots & & \vdots & \vdots & \vdots & & \vdots \\ -\alpha_{m,1} & -\alpha_{m,2} & \cdots & -\alpha_{m,n} & 0 & 0 & \cdots & 0 \end{bmatrix} \quad \text{(A3-4)}$$

Given that $\mathbf{x}$ is a solution to $\partial L/\partial \mathbf{x} = \mathbf{0}$, as in Equation (A3-3), a sufficient condition for minimization of Equations (7-5) is that Equation (A3-4) be *positive definite* (Selby 1969) for that solution:

$$\mathbf{x}'\left(\partial(\partial L/\partial \mathbf{x})/\partial \mathbf{x}'\right)\mathbf{x} > \mathbf{0} \quad \text{(A3-5)}$$

A sufficient condition for maximization is

$$\mathbf{x}'\left(-\partial(\partial L/\partial \mathbf{x})/\partial \mathbf{x}'\right)\mathbf{x} > \mathbf{0} \quad \text{(A3-6)}$$

By using Equation (A3-4),

$$\mathbf{x}'\left(\partial(\partial L/\partial \mathbf{x})/\partial \mathbf{x}'\right)\mathbf{x} = 2\sum_{i=1}^{n} w_i^2 - \sum_{i=1}^{n} w_i \sum_{k=1}^{m} \lambda_k \alpha_{k,i} - \sum_{k=1}^{m} \lambda_k \sum_{i=1}^{n} \alpha_{k,i} w_i$$

$$= 2\sum_{i=1}^{n} w_i^2 - 2\sum_{i=1}^{n} w_i \sum_{k=1}^{m} \lambda_k \alpha_{k,i} \quad \text{(A3-7)}$$

However, by Equation (A3-3), $\sum_{k=1}^{m} \lambda_k \alpha_{k,i} = 2(w_i - 1)$ and Equation (A3-7) becomes

$$\mathbf{x}'\left(\partial(\partial L/\partial \mathbf{x})/\partial \mathbf{x}'\right)\mathbf{x} = 2\sum_{i=1}^{n} w_i^2 - 2\sum_{i=1}^{n} w_i 2(w_i - 1)$$

$$= 4\sum_{i=1}^{n} w_i - 2\sum_{i=1}^{n} w_i^2 \quad \text{(A3-8)}$$

Note also that because Equations (7-5b) always includes Equation (6-3), in which the sum of the n weights equals n, the first sum in the last line of Equation (A3-8) can be replaced with n. Therefore Equations (A3-5), (A3-8), and (6-3) imply that solution $\mathbf{x}$ represents a minimum if

$$\sum_{i=1}^{n} w_i^2 < 2n \quad \text{(A3-9)}$$

Likewise, Equations (A3-6), (A3-8), and (6-3) imply a maximum if

$$\sum_{i=1}^{n} w_i^2 > 2n \quad \text{(A3-10)}$$

194

SUFFICIENCY FOR EQUATIONS (8-17)

Likewise, recall that the Lagrangian for the problem of Equations (8-11) is

$$L = \sum_{i=1}^{n}(w_i - 1)^2 - \sum_{k=1}^{M}\lambda_k\left(\sum_{i=1}^{N}\alpha_{k,i}w_i - e_k\right) \tag{A3-11}$$

where $M = m + p + q$ and $N = n + p + q$. Now let

$$\mathbf{x}' = \begin{bmatrix} w_1 & \cdots & w_n & w_{n+1} & \cdots & w_N & \lambda_1 & \cdots & \lambda_M \end{bmatrix} \tag{A3-12}$$

and let $\mathbf{x}$ be defined similarly as the column vector. Equations (8-17) are written in vector notation as

$$\frac{\partial L}{\partial \mathbf{x}} = \begin{bmatrix} \partial L/\partial w_1 \\ \vdots \\ \partial L/\partial w_n \\ \partial L/\partial w_{n+1} \\ \vdots \\ \partial L/\partial w_N \\ \partial L/\partial \lambda_1 \\ \vdots \\ \partial L/\partial \lambda_M \end{bmatrix} = \begin{bmatrix} 2(w_1 - 1) - \sum_{k=1}^{M}\lambda_k\alpha_{k,1} \\ \vdots \\ 2(w_n - 1) - \sum_{k=1}^{M}\lambda_k\alpha_{k,n} \\ -\sum_{k=1}^{M}\lambda_k\alpha_{k,n+1} \\ \vdots \\ -\sum_{k=1}^{M}\lambda_k\alpha_{k,N} \\ -\sum_{i=1}^{N}\alpha_{1,i}w_i + e_1 \\ \vdots \\ -\sum_{i=1}^{N}\alpha_{M,i}w_i + e_M \end{bmatrix} = 0 \tag{A3-13}$$

The second derivative matrix is

$$\frac{\partial(\partial L/\partial \mathbf{x})}{\partial \mathbf{x}'} = \begin{bmatrix} 2 & \cdots & 0 & 0 & \cdots & 0 & -\alpha_{1,1} & \cdots & -\alpha_{M,1} \\ \vdots & & \vdots & \vdots & & \vdots & \vdots & & \vdots \\ 0 & \cdots & 2 & 0 & \cdots & 0 & -\alpha_{1,n} & \cdots & -\alpha_{M,n} \\ 0 & \cdots & 0 & 0 & \cdots & 0 & -\alpha_{1,n+1} & \cdots & -\alpha_{M,n+1} \\ \vdots & & \vdots & \vdots & & \vdots & \vdots & & \vdots \\ 0 & \cdots & 0 & 0 & \cdots & 0 & -\alpha_{1,N} & \cdots & -\alpha_{M,N} \\ -\alpha_{1,1} & \cdots & -\alpha_{1,n} & -\alpha_{1,n+1} & \cdots & -\alpha_{1,N} & 0 & \cdots & 0 \\ \vdots & & \vdots & \vdots & & \vdots & \vdots & & \vdots \\ -\alpha_{M,1} & \cdots & -\alpha_{M,n} & -\alpha_{M,n+1} & \cdots & -\alpha_{M,N} & 0 & \cdots & 0 \end{bmatrix} \tag{A3-14}$$

By using Equation (A3-14),

$$\mathbf{x}'\big(\partial(\partial L/\partial\mathbf{x})/\partial\mathbf{x}'\big)\mathbf{x} \;=\; 2\sum_{i=1}^{n} w_i^2 \;-\; \sum_{i=1}^{n} w_i \sum_{k=1}^{M} \lambda_k \alpha_{k,i} \;-\; \sum_{i=n+1}^{N} w_i \sum_{k=1}^{M} \lambda_k \alpha_{k,i}$$

$$-\; \sum_{k=1}^{M} \lambda_k \sum_{i=1}^{n} \alpha_{k,i} w_i \;-\; \sum_{k=1}^{M} \lambda_k \sum_{i=n+1}^{N} \alpha_{k,i} w_i \qquad \text{(A3-15)}$$

$$=\; 2\sum_{i=1}^{n} w_i^2 \;-\; 2\sum_{i=1}^{n} w_i \sum_{k=1}^{M} \lambda_k \alpha_{k,i} \;-\; 2\sum_{i=n+1}^{N} w_i \sum_{k=1}^{M} \lambda_k \alpha_{k,i}$$

However, by Equation (A3-13), $\displaystyle\sum_{k=1}^{M} \lambda_k \alpha_{k,i} = 2(w_i - 1)$, $i = 1, \ldots, n$, and $\displaystyle\sum_{k=1}^{M} \lambda_k \alpha_{k,i}$
$= 0$, $i = n+1, \ldots, N$, and Equation (A3-15) becomes

$$\mathbf{x}'\big(\partial(\partial L/\partial\mathbf{x})/\partial\mathbf{x}'\big)\mathbf{x} \;=\; 2\sum_{i=1}^{n} w_i^2 \;-\; 2\sum_{i=1}^{n} w_i 2(w_i - 1)$$

$$=\; 4\sum_{i=1}^{n} w_i \;-\; 2\sum_{i=1}^{n} w_i^2 \qquad \text{(A3-16)}$$

Therefore Equations (A3-5), (A3-16), and (6-3) imply that solution $\mathbf{x}$ represents a minimum if

$$\sum_{i=1}^{n} w_i^2 \;<\; 2n \qquad \text{(A3-17)}$$

Likewise, Equations (A3-6), (A3-16), and (6-3) imply a maximum if

$$\sum_{i=1}^{n} w_i^2 \;>\; 2n \qquad \text{(A3-18)}$$

Equations (A3-17) and (A3-18) are the same as Equations (A3-9) and (A3-10), so the same sufficiency condition exists for both Equations (7-7) and (8-17).

REFERENCES

Barnston, A. G., and C. F. Ropelewski (1992). "Prediction of ENSO episodes using canonical correlation analysis." *Journal of Climate*, AMS, **5**(11):1316–45.

Barnston, A. G., H. M. van den Dool, D. R. Rodenhuis, C. F. Ropelewski, V. E. Kousky, E. A. O'Lenic, R. E. Livezey, M. Ji, A. Leetmaa, S. E. Zebiak, M. A. Cane, T. P. Barnett, and N. E. Graham (1994). "Long-lead seasonal forecasts—Where do we stand?" *Collected Papers on Seasonal Forecasting for the Workshop on Long-Lead Climate Forecasts,* Chicago, November 4, National Oceanic and Atmospheric Administration, 40 pp.

Chow, V. T. (1964). *Handbook of Applied Hydrology.* McGraw-Hill, New York, 8–29.

Croley, T. E., II (1993). "Probabilistic Great Lakes hydrology outlooks." *Water Resources Bulletin*, AWRA, **29**(5):741–53.

Croley, T. E., II (1996). "Using NOAA's new climate outlooks in operational hydrology." *Journal of Hydrologic Engineering* **1**(3):93–102.

Croley, T. E., II (1997a). "Water resource predictions from meteorology probability forecasts." *Proceedings of the Symposium on Sustainability of Water Resources under Increasing Uncertainty*, Rabat, Morrocco, D. Rosbjerg et al., ed., IAHS Publication No. 240, IAHS Press, Institute of Hydrology, Wallingford, Oxfordshire, 301–09.

Croley, T. E., II (1997b). "Mixing probabilistic meteorology outlooks in operational hydrology." *Journal of Hydrologic Engineering* **2**(4):161–68.

Croley, T. E., II (1998). "Great Lakes Advanced Hydrologic Prediction System." *Proceedings of the First Federal Interagency Hydrologic Modeling Conference*, Federal Subcommittee on Hydrology of the Interagency Advisory Committee on Water Data, Water Information Coordination Program, US Geological Survey, Reston, Virginia, **2**:6-1–6-8.

Croley, T. E., II, and H. C. Hartmann (1990). "GLERL's near real-time hydrology outlook package." *Proceedings of the Great Lakes Water Level Forecasting and Statistics Symposium*, Great Lakes Commission, Windsor, Ontario, May 17–18, 63–72.

Croley, T. E., II, and D. H. Lee (1993). "Evaluation of Great Lakes net basin supply forecasts." *Water Resources Bulletin*, AWRA, **29**(2):267–82.

Croley, T. E., II, F. H. Quinn, K. E. Kunkel, and S. A. Changnon (1996). "Climate transposition effects on the Great Lakes hydrology cycle." *NOAA Technical Memorandum ERL GLERL-89*, Great Lakes Environmental Research Laboratory, Ann Arbor, Michigan, 100 pp.

Day, G. N. (1985). "Extended streamflow forecasting using NWSRFS." *Journal of the Water Resources Planning and Management Division*, ASCE, **111**:157–70.

Epstein, E. S. (1988). "Long-range weather prediction: Limits of predictability and beyond." *Weather and Forecasting* **3**(1):69–75.

Gilman, D. L. (1985). "Long-range forecasting: The present and the future." *Bulletin of the American Meteorology Society*, AMS, **66**(2):159–64.

Halpert, M. S., and C. F. Ropelewski (1992). "Surface temperature patterns associated with the Southern Oscillation." *Journal of Climate* **5**:577–93.

Hillier, F. S., and G. J. Lieberman (1969). *Introduction to Operations Research.* Holden-Day, San Francisco. Chapter 5: Linear Programming, 127–71. Also Appendix 2: Classical Optimization Techniques, 603–08.

Huang, J., H. M. van den Dool, and A. G. Barnston (1994). "Long-lead seasonal temperature prediction using optimal climate normals." *Collected Papers on Seasonal Forecasting for the Workshop on Long-Lead Forecasts,* Chicago, November 4, National Oceanic and Atmospheric Administration, 16 pp.

Ingram, J. J., M. D. Hudlow, and D. L. Fread (1995). "Hydrometeorology coupling for extended streamflow predictions." *Preprints of Conference on Hydrology,* 75th Annual Meeting of the American Meteorology Society, Dallas,Texas, January 15–20, 186–91.

Ji, M., A. Kumar, and A. Leetmaa (1994). "A multiseason climate forecast system at the National Meteorology Center." *Bulletin of the American Meteorology Society,* AMS, **75**(4):569–77.

Linsley, R. K, M. A. Kohler, and J. L. H. Paulhus (1958). *Hydrology for Engineers.* McGraw-Hill, New York, 193–215.

Livezey, R. E. (1990). "Variability of skill of long-range forecasts and implications for their use and value." *Bulletin of the American Meteorology Society,* AMS, **71**(3):300–09.

Pfieffer, P. E. (1965). *Concepts of Probability Theory.* McGraw-Hill, New York.

Rasmusson, E. M. (1984). "El Niño: The ocean/atmosphere connection." *Oceanus* **27**(2):5–12.

Ritchie, H. (1991). "Application of the semi-Lagrangian method to a multi-level spectral primitive-equations model." *Quarterly Journal of the Royal Meteorology Society* **117**:91–106.

Ritchie, H., C. Beaudoin, and A.-M. Leduc (1995). "An upgrade of the Canadian global spectral forecast model." *Canadian Meteorology Centre Review* **2**(3):2–17.

Ropelewski, C. F., and M. S. Halpert (1986). "North American precipitation and temperature patterns associated with the El Niño/Southern Oscillation (ENSO)." *Monthly Weather Review* **114**(12):2352–62.

Ropelewski, C. F., and P. D. Jones (1987). "An extension of the Tahiti-Darwin Southern Oscillation index." *Monthly Weather Review* **115**:2161–65.

Selby, S. M. (1969). *CRC Standard Mathematical Tables.* The Chemical Rubber Co., Cleveland, 131–37.

Shabbar, A., B. Bonsal, and M. Khandekar (1997). "Canadian precipitation patterns associated with the Southern Oscillation." *Journal of Climate* **10**:3016–27.

Shabbar, A., and M. Khandekar (1996). "The impact of El Niño–Southern Oscillation on the temperature field over Canada." *Atmosphere Ocean* **34**(2):401–16.

Smith, J. A., G. N. Day, and M. D. Kane (1992). "Nonparametric framework for long-range streamflow forecasting." *Journal of Water Resources Planning and Management,* ASCE, **118**(1):82–92.

van den Dool, H. M. (1994). "Long-range weather forecasts through numerical and empirical methods." *Dynamics of Atmospheres and Oceans* **20**:247–70.

Wagner, A. J. (1989). "Medium- and long-range forecasting." *Weather and Forecasting* **4**(3):413–26.

INDEX

DATE DUE

HIGHSMITH #45230

Printed
in USA